THE SINGING GORILLA

THE SINGING GORILLA

Understanding Animal Intelligence

George Page

HEADLINE

First published in 1999
by HEADLINE BOOK PUBLISHING

By arrangement with the BBC

10 9 8 7 6 5 4 3 2 1

British Library Cataloguing in Publication Data

Page, Gregory
The singing gorilla
1.Animal intelligence
I.Title
591.5'13

ISBN 0 7472 7568 8 (hardback)
ISBN 0 7472 7569 6 (softback)

The Publishers would like to thank the following
for providing photographs:

BBC NHU/Tony Phelps page 7t; The Bridgeman Art Library 1t;
Bruce Coleman Collection/Alain Compost 4b, Christer Fredriksson 8;
Mary Evans Picture Library 1b; Lizzie Green 2t, 3, 7b;
Andrew Murray 2b, 6t; OSF/F Koester, Survival Anglis 4t;
Andrew Plumptre 5; Clive Bromhall 6b.

Typeset by Palimpsest Book Production Limited,
Polmont, Stirlingshire
Printed and bound in Great Britain by
Mackays of Chatham PLC, Chatham, Kent

HEADLINE BOOK PUBLISHING
A division of Hodder Headline PLC
338 Euston Road
London NW1 3BH

For Dr Dennis De Stefano

'We must understand what to make of our encounter with the animals. Because as we ourselves prosper in unseemly numbers they vanish, and in the end our prosperity may amount to nothing without them.'

Paul Shepard, *The Others*, 1996

CONTENTS

PREFACE

As executive editor, host and narrator of the US public television series, *Nature*, I have spent many wonderful hours and days observing animals in the wild. A quick count puts me in about 20 countries on exactly five continents over almost two decades. Like many travellers, I found East Africa the most compelling landscape on earth. There is no place like the Masai Mara for getting a real sense of the extremely complex lives of elephants, lions, leopards, hyenas, wildebeest, monkeys, wild dogs, and many more – the whole rich panoply. Back in our surburban home, north of New York City, I have spent hundreds of equally enjoyable hours playing with and caring for my three dogs and one cat. And, like anyone who has ever observed a stalking lioness, loved a pet, watched a bird fly hither and yon in search of just the right nesting material, or held the gaze of a chimpanzee, I've often wondered, 'Are these creatures "thinking" all the while? Without language as we know it, *how* are they thinking?'

For almost 100 years, the behaviourist tradition of psychology has argued that in fact these animals have no 'minds' at all. According to this tradition, any belief to the contrary is 'folk psychology' – mere anthropomorphism, whereby we ascribe human characteristics to animals (or even to inanimate objects). Since we cannot 'get inside' the animal's mind (as we can get inside our own minds), and since the animal cannot report what's going on – not in a 'language' we can understand – all we have left are guesses and speculation fatally tainted by anthropomorphism, with no provable facts. But does the reductive hypothesis of the behaviourists make any sense outside the laboratory? After all, there is no way to prove conclusively what is going on in another human being's mind. Can it possibly be true that I am the only conscious creature on this planet? Not likely!

What are animals feeling? Homer said that Ulysses' dog Argos recognised his master after an absence of 20 years. The *Odyssey* is a fable, of course, but we can readily imagine that a dog could have such a keen memory of his master. Indeed, our modern-day pets behave in ways that we feel compelled to interpret in terms of our own emotions. Elephants in the wild certainly seem to demonstrate sadness, even depression, at appropriate turning points in their lives, as well as behaviour that appears to be true joy. Are the dog and the elephant feeling what we would feel in similar situations? If we could exchange emotional experiences with either animal, would we feel quite at home? Again, the behaviourist tradition scoffs at the idea.

One final question: what difference does it make? Well, what difference does it make how far it is to the end of the universe? Or if there is an 'end' to the universe? What difference does it make if quarks are the building blocks of the universe, or if there exist entities even tinier and more fundamental? We humans are as curious as cats; we want to know. Even more important, our ideas about animal minds and emotions influence far-reaching questions of ethics and politics. If we believe or, better yet, determine that animals think consciously and possess the full range of 'human' emotions – if their pain is therefore also 'suffering' – on what ethical grounds do we support, for example, marine theme parks, whose great performing whales and dolphins are, in the final analysis, enslaved in what to them are tiny 'puddles'? We become hard-pressed to condone animal experimentation and a growing number of people decide that they cannot justify eating meat. On the other hand, some ethicists fear that not maintaining a distinction between ourselves and all other animals, including the porpoises and great apes, might set us on a slippery slope that could lead to all manner of abuses of human beings – the same abuses to which we now subject these other creatures, on the grounds that they are fundamentally different from us. I do not share this fear.

Personally, I very much like the idea that my kinship with the other living creatures of the earth might be much greater than our cultural and scientific heritage has led us to believe. I am comfortable with the fact that, like them, I am an animal and that many of the characteristics that I have so often tossed off as being just part of 'human nature', may in fact be aspects of our shared animal natures.

Or, as Cynthia Moss, who has studied elephants in East Africa for more than 30 years, put it rather better: 'We are just part of a continuum of species. It isn't just animals and man, I think that's becoming clear. We are just another much more intelligent and masterful animal but we are all part of this earth.'

So, questions about animal minds are not merely fascinating. The answers matter.

chapter 1

FROM ALTAMIRA TO ANTHROPOMORPHISM

Ancient Heritage

In the beginning, our lives were totally immersed in the world of animals. In the beginning, in fact, we *were* animals, in the colloquial as well as technical sense of the word. Our Neanderthal and Cro-Magnon ancestors relied on birds and beasts for food and clothing; they engaged in single combat with tigers and bison and bears, and they took heavy casualties. We were an integral part of a world that was red in tooth and claw.

Surely it is no coincidence that our oldest surviving visual art (the cave paintings at Altamira, Lascaux and other sites in southern Europe) depicts mainly cattle, horses, bison and deer. Whether shown as hunters' prey or as objects of veneration, animals predominate; there are very few human figures in any of this work. And our earliest music may well have been created in response to the myriad sounds of the natural world. After all, those were the only sounds we heard: there weren't any jets, drills or jukeboxes around, but there were songbirds in profusion. No wonder Orpheus, the mythical Greek musician, could mesmerise a rapt menagerie of wild beasts with every note of his lyre.

While the very earliest Paleolithic art is dated about 30,000 years ago, the famous work in southern Europe goes back to about 10,000 BC. Anthropologists believe that those highly artistic ancestors were still exclusively hunters and gatherers, but their horizons would soon expand. At about the same time as the Solutreo-Magdalenian artists were producing their greatest 'canvases' at Altamira, the peoples of Mesopotamia and the Middle East were beginning to domesticate mammals. In the long story of our relationship with animals, and our still-evolving understanding of the animal mind, domestication marked the beginning of our estrangement. It was a watershed of incalculable importance – small wonder that it was incorporated into all the creation stories of the time.

The first chapter of Genesis says:

> And God made the beast of the earth after his kind, and cattle after their kind, and every thing that creepeth upon the earth after his kind: and God saw that it was good. And God said, 'Let us make man in our image, after our likeness: and let them have dominion over the fish of the sea, and over the fowl of the air, and over the earth, and over every creeping thing that creepeth upon the earth.' So God created man in his own image, in the image of God he created them; male and female he created them. And God blessed them, and God said unto them, 'Be fruitful, and multiply, and replenish the earth, and subdue it; and have dominion over the fish of the sea, and over the fowl of the air, and over every living thing that moveth upon the earth.'

The subsequent thirtieth verse of that chapter reads: 'Also, to every beast of the earth, to every bird of the air, and to everything that creeps on the earth, in which there is life, I have given every green herb for food; and so it was.'

Some commentators believe this verse implies that the Garden of Eden before the Fall was a vegetarian society. I diplomatically take no position on this provocative notion, but there can be no doubt about the disposition of the earth's resources following the Flood:

> So God blessed Noah and his son and said unto them, 'Be

fruitful, and multiply, and replenish the earth. And the fear of you and the dread of you shall be upon every beast of the earth, and upon every fowl of the air, upon all that moveth upon the earth, and upon all the fishes of the sea; into your hands are they delivered. Every moving thing that liveth shall be meat for you; even as the green herb have I given you all things.'

In short, God gives mankind dominion over *all* that lives on this earth. Various passages throughout the Old Testament prescribe respectful treatment for domesticated and wild animals and the same holds for Judaic Law, but the essential message of the Pentateuch and the legacy of the Old Testament are clear enough: animals were created, or designated, for our needs. Nowhere in the New Testament is this message challenged.

In Matthew 6:26, Jesus says in the Sermon on the Mount, in the verses that precede the beautiful 'lilies of the field' image: 'Look at the birds of the air, that they do not sow, neither do they reap, nor gather into barns, and yet your heavenly Father feeds them. Are you not worth much more than they?' In his first Letter to the Corinthians, the apostle Paul reiterates this theme when he writes, 'For it is written in the law of Moses, "You shall not muzzle an ox while it treads out the grain." ' Is it oxen God is concerned about? Or does He say it altogether for our sakes? For our sakes, no doubt it is written. And, writing 350 years later, St Augustine cites Jesus's withering of the fig tree which had failed to bear fruit, and his sending of the devils into the herd of swine, as proof that our actions in relation to nature are not subject to the moral code that should govern our dealings with other men and women. The natural world is our rightful domain.

Nor could this relationship between man and animal be otherwise in the West because it flows necessarily from our having been created by God in His image. If humans are kin to God, aren't we therefore fundamentally different from every animal? As God has dominion over us, so must we have dominion over them. It is logical, and it is written. (This, incidentally, partly explains the adamant prohibition against bestiality in Western culture. Unlike incest, also the subject of the strictest censure, bestiality is an act without bad reproductive consequences, so to speak, but it does lower the status of man by moving him away from God and towards the fallen world. It is the gravest insult to the Creator, and thus it is not tolerated.)

I should mention here that I was raised a Methodist in Georgia. I recall these biblical passages not in order to challenge the Christian faith in any way, but simply to underline the heritage of our prevailing belief system regarding the nature and status of animals. Besides, our other major cultural influence, the philosophical tradition originating in ancient Greece, was also adamant about the status of animals. Some Greek philosophers did hypothesise something very like a 'mind' for wild beasts, but Aristotle was the most important of the philosophers in this field, and for Aristotle man was the only creature possessed of reason, consciousness and a soul.

According to the indispensable *Oxford Companion to Animal Behaviour*, the great thinker discussed, to one degree or another, 540 different species of animals in his writings. He was the first great naturalist, and his ideas about the natural world were the most important from any source for fully 1500 years. As we shall see, Aristotle had some modern-sounding ideas on this subject, but he could also be amazingly uninformed and naive. He apparently accepted at face value the fable that the mother bear literally licks her progeny into the proper bear-like shape. Certainly the philosopher who rationalised human slavery had no concept of 'animal rights' as entertained by many of us over two millennia later.

The Romans, too, were fascinated by animals, and their wealth and vast empire allowed them to satisfy their curiosity by capturing huge numbers of wild beasts. The emperor Augustus was said to have had a collection of 420 tigers; Nero, 400 bears; Trajan, 11,000 animals of practically every known species. Pliny the Elder was a first-century Roman civil servant and naturalist whose 37-volume *Historia Naturalis* was packed with almost every imaginable fantasy and falsehood about the natural world — and a few facts as well.

Since he was writing 400 years after Aristotle, it is surprising to find that Pliny had the same bizarre faith in the creative powers of the mother bear. His belief in spontaneous generation – as the explanation for the sudden appearance of maggots on a carcass – was logical enough in that age. But what are we to make of the report that a distant kingdom in Africa was ruled by a dog-headed race of people? Perfectly believable as well, apparently, because there it is in Pliny!

Change came painfully slowly in those days. In fact, Western culture changed very little over 1000 years, so it's not surprising that Christian monks living in the twelfth and thirteenth centuries AD were still highly susceptible to such fantastical stories, which they found in a third-century text that picked up where Pliny had left off. This was the *Physiologus*, or *The Natural Historian*, and these animal stories provided a wonderful subject for medieval illustrators. The resulting volumes, now known as 'bestiaries', also drew on the residue of the various pagan religions which had been absorbed as Roman Christianity spread north, south, east and west. Paganism – always, everywhere, and in every form – consecrates animal powers and animal mysteries.

In the twelfth and thirteenth centuries, it did not occur to most people to hike up a mountain (which was seen literally as another world) or to walk on the seashore (the boundary of yet another demonstrably dangerous world). Just as few ventured forth into the field to study natural history at first hand. Thus, for several centuries, the stories and images in these bestiaries comprised the sum total of popular knowledge of the animal kingdom. Inevitably, the bestiaries became more and more fanciful, as editors added new and stranger stories, and the gold-leaf illuminations became more spectacular. Several hundred of these gorgeous books have survived intact, and they include wonderful depictions of fire-breathing dragons and lovely mermaids.

By and large, pets were not as revered in the Middle Ages as they are today. To keep one could be positively dangerous, in fact, leading one to be suspected of witchcraft. Hard as it is for us to believe today, animals were put on criminal trial throughout these centuries. This bizarre cultural phenomenon is reported by Dorothy Cheney and Robert Seyfarth in their excellent and important book, *How Monkeys See the World*. Cheney and Seyfarth quote from *The Criminal Prosecution and Capital Punishment of Animals*, a volume by one E.P. Evans, first published in 1906 and currently out of print, alas. The cited story details the charges against three sows who had been feeding quietly until, 'excited and enraged by the squealing of one of the porklings', they trampled the son of the swinekeeper, who died some days later. 'After due process of law,' the three sows were condemned to death, and so were their peers in the herd, who had 'hastened to the scene of the murder and by their cries and aggressive actions showed that they approved of the assault, and were ready and even eager to become *participes criminis*.'

In contrast, various famous Christians down the ages are remembered for their kindness towards animals – St Francis of Assisi especially – but, by and large, the attitude of the mainstream Church towards the animal kingdom had not changed much from biblical days. The message of the Scriptures and the hierarchy of the universe were plain for all to see. Thomas Aquinas, the great thirteenth-century churchman who reconciled Catholic teachings with the recently rediscovered philosophy of Aristotle, understood that animals could suffer, but he based his opposition to cruelty to them only upon the supposition that this cruelty might lead in turn to cruelty to humans or might rebound as an economic loss to the perpetrator. In his great *Summa Theologiae*, we are the only creatures who really count. The universe is still 100 per cent anthropocentric.

Moving into the Modern: Descartes to Darwin

But surely ideas about animals changed in the Renaissance? Well, yes and no. Leonardo da Vinci was a vegetarian; Montaigne wrote that cruelty to animals was wrong in and of itself; and the tradition of careful observation of the real world gained credence. Even Thomas

More, the justly esteemed English statesman and Catholic martyr, digressed in his *Utopia* to describe quite accurately the phenomenon of imprinting-at-birth in barnyard fowl (newly hatched chicks will imprint on the first creature they see – human or beast – and will consider that creature 'mum', which fortunately it usually is), almost 400 years before the definitive work of Konrad Lorenz in the twentieth century. On the other hand, we have René Descartes.

Most discussions of Western ideas about animal consciousness completely skip over the ancient Greeks and Romans and the first 1500 years of Christian thought. Instead they proceed directly to this pivotal French philosopher and mathematician, who lived and wrote in the first half of the seventeenth century. Descartes was adamant in his belief that only human beings are endowed with consciousness. Animals need not apply. Furthermore, Descartes argued that animals were nothing more or less than pieces of soft, fleshy machinery, albeit quite complicated in many cases. He was forced to take this extreme mechanistic position because he had backed himself into a corner in his defence of the immortal human soul, about which he was also adamant, as was everyone else at that time.

Descartes' logic goes something like this: human beings have souls, as taught by Christian doctrine and as distinguished by consciousness; animals do not have souls because they cannot have souls, according to Christian doctrine; therefore animals cannot have consciousness either. (One of the Early Church's most ingenious arguments against animals having souls was that neither heaven nor hell would be big enough to contain the souls of all humans *and* animals that ever lived.)

This is a simplified version of Descartes' thinking, and I am no logician, but I still sense a lot of holes in his argument. And I take comfort from the fact that philosophers have found it problematic too. I also think of Benjamin Franklin's famous, irreverent quip that the advantage of our being 'reasonable creatures' is that we can always come up with a reason for doing and thinking whatever we want to. Descartes was a brilliant man, sometimes referred to as the father of modern philosophy – *Cogito, ergo sum* and all that – and he made major contributions in optics and mathematics. But, as the single most important figure in the story of Western attitudes to animal consciousness, his contribution is painfully negative and, as I think we have discovered in the twentieth century, dead wrong.

But Descartes was not alone in taking this rigidly 'mechanistic' approach. Any number of other brilliant thinkers before and after him subscribed to similar views of animal nature. The Englishman Thomas Hobbes was a thoroughgoing materialist with a pessimistic worldview that produced his infamous depiction of human life (let alone the lives of animals) as 'nasty, brutish, and short'. Another famous Englishman, Descartes' near-contemporary Francis Bacon, presupposed a mechanistic world suitable for human domination as the basis for scientific empiricism. Surely it is no coincidence that all manner of cruel experimentation on animals became more and more common in the seventeenth century, as these Cartesian ideas were disseminated across Europe.

The following appalled and appalling description comes to us from a contemporary of Descartes, the Enlightenment free thinker Voltaire, writing 100 years later:

> They administered beatings to dogs with perfect indifference, and made fun of those who pitied the creatures as if they felt pain. They said the animals were clocks; that the cries they emitted when struck were only the noise of a little spring that had been touched, but that the whole body was without feeling. They nailed poor animals up on boards by their four paws to vivisect them and see the circulation of the blood which was a great subject of conversation.

Ironically, however, the results of these grotesque anatomy lessons only compelled more and more people to question the morality of such experiments, because they demonstrated the essential sameness of physiology among all mammals, including man.

Voltaire, again:

> . . . Judge this dog who has lost his master, who has searched

for him with mournful cries in every path, who comes home agitated, restless, who runs up and down the stairs, who goes from room to room, who at last finds his beloved master in his study, and shows him his joy by the tenderness of cries, by his leaps, by his caresses. There are barbarians who seize this dog, who so greatly surpasses man in fidelity and friendship, and nail him down to a table and dissect him alive, to show you the mesenteric veins. You discover in him all the same organs of feeling as in yourself. Answer me, mechanist, has Nature arranged all the springs of feeling in this animal to the end that he might not feel? Does he have nerves [in order] to be impassive?

David Hume, a contemporary of Voltaire, seconded this pointed sarcasm and added his own: 'Next to the ridicule of denying an evident truth, is taking such pains to defend it; and no truth appears to me more evident, than that beasts are endowed with thought and reason as well as man.'

Hume and Voltaire and their sympathisers employed what philosophers call 'the argument from inference'. In other words, just as we laypeople infer that other humans feel pain and suffering based on our own experience of pain and suffering, so we must infer the same of other vertebrates, because the vertebrate nervous system is essentially identical to our own.

A century later, Charles Darwin extended this argument from inference about as far as it could go. The scientist who was – and still is, in some circles – reviled as the man who 'lowered' mankind to the level of the animal (with his theory of evolution and natural selection) can also be hailed as the man who did his utmost to raise animals as near as possible to the level of humans. Darwin's ideas presupposed a continuum of consciousness in the natural world, from the 'lowest' invertebrate to the 'highest' primate (us). Neither the human mind nor the human emotions could have appeared suddenly and without precedent. They must have evolved from 'incipient' predecessors.

The Anthropomorphic Instinct

What an intriguing turns of events in Western intellectual history! In the seventeenth century, René Descartes had felt it necessary to deny consciousness to animals as part of a philosophical project to guarantee the existence of the human soul. Now, in the nineteenth century, Charles Darwin had ulterior motives for defending animal consciousness. At the heart of the apparent conflict between these two positions was – and still is – the age-old issue of anthropomorphism. In its simplest terms, this means the use of human characteristics to describe the non-human; or, to put it another way, ascribing human characteristics to non-humans. To shout at a stalled car for being stubborn is anthropomorphising, because cars aren't stubborn. Nor is yon lazy cloud really lazy, nor the angry hurricane truly angry as it churns out of the Caribbean and into the Gulf of Mexico.

This is just the way we enjoy thinking and talking. *Why* we tend to relate to the natural world in this way is a fascinating question in its own right, the subject of numerous books and studies, popular and academic. It is also vital to our discussion, because many scientists adopt some version of the Cartesian position – that to believe that animals are thinking, conscious beings is merely anthropomorphising. From their perspective, a dog – even one with its four paws 'planted in concrete', refusing to budge – is no more stubborn than the stalled car. That adjective is an unwarranted assumption on our part in each case.

I strongly disagree, but the naysayers definitely have one valid point. It's true that in the formative years of every human culture, we were obsessed with animals. One of the earliest writers to appreciate this anthropomorphic tendency was Xenophanes, a Greek thinker of the sixth century BC. He wrote, 'But if oxen and horses . . . could create works of art like those made by men, horses would draw pictures of gods like horses, and oxen of gods like oxen.' He's probably right about that. Furthermore, in the formative years of our individual lives, many if not most of us are infatuated with animals. In fact, researchers have found that children dream mainly about animals (as we grow into adulthood, the self supersedes them as our favourite dream subject).

I was nine years old when I got my first dog. She was a cuddly little cur about four months old, part collie and part sheepdog. In a burst of great originality, about which I was not the least bit embarrassed, I named her Lassie. It was love at first sight for both of us and we became inseparable. Lassie would follow me to school and wait for me patiently until school finished. We had to put a stop to this after a few days, however, because it was not safe for her to be running around off the leash even in our small Georgia town.

Lassie slept with me, of course, and would get up at the crack of dawn with my parents so that she could go outside. When it was time for me to get up a little later, Mother would say to Lassie, 'Go get George!' Lassie would bound up the stairs at the speed of light and pounce on me with her full weight. If I did not get up immediately, she would pull the covers off.

Did I think Lassie was a thinking, emotional and conscious being? To the extent that I thought about such things at all at that age, I most certainly did. Like all children – and adults – who fall in love with an animal, I was completely anthropomorphic towards Lassie. I would talk to her endlessly, divulging my deepest childhood secrets, and I had no doubt whatsoever that she understood every word I said, even if her reaction was just a little 'woof' and a snuggle to get closer to me. I know now, of course, that, while Lassie did not understand every word I said to her, she deftly interpreted my body language and the tone of my voice to read my moods. In fact, I have not known a dog since who was so good at making me believe that I was being completely understood. All dogs possess this ability to varying degrees, but Lassie was the champ.

Lassie and I shared a wonderful life together for three years. And then one summer during the war my parents decided we were going to take a rare four-week vacation and, for the first time ever, we would just close down our big old house and let the housekeeper check on it occasionally. I was delighted with the news and said, 'Oh boy, Lassie is going to love this trip!'

'Well,' said my father, 'your mother and I have given the question of Lassie considerable thought and we've decided that Lassie would be happier if we boarded her at our vet's very nice kennel. After all, we're going to be driving a lot and Lassie does have a tendency to get car-sick.'

I was crestfallen and fought the urge to tell my parents that I would just not go on the trip if I could not take Lassie. Upon reflection, however, I realised that no amount of protesting from me – especially in the form of threats – was going to change my parents' minds so, reluctantly, I gave in. That night I dreamed that Lassie and I ran away. But, when we got hopelessly lost somewhere in China, the dream of liberation turned into a nightmare. The next day my mother and I took Lassie to the kennel. I tried very hard to 'be a man' but, as we drove away after leaving Lassie, I could not hold back the tears. I looked over at my mother and she was crying, too.

We went to Florida and literally toured the state from top to bottom. I had never been there before and I loved every minute of it. For the first few days I called the kennel every day to check on Lassie. I called on the Monday of our last week away and, since we would be home the following Friday, I did not call any more, knowing that Lassie was fine.

On Thursday of that week my mother called home from Tampa to check on things while I was cavorting in the hotel's swimming pool. When I came upstairs later I sensed immediately that something was wrong. My mother's eyes were red and my father sat in a chair looking completely dejected. Mom gestured for me to come to her and she put her arms around me. 'Son,' she said, 'I got some bad news when I called home.'

I stiffened my body and then heard her say in a cracked voice, almost a whisper, 'Lassie died last night.' And then all three of us just cried and cried. I had never seen my father cry before.

The vet said his kennel man had checked on Lassie and the other boarders at midnight and Lassie was fine. When he made his rounds at 3 a.m., Lassie was dead. The vet told us he had absolutely no idea why my dog had died. He even performed an autopsy and found only that her healthy heart had just stopped beating. 'We never had anything like this happen before,' he said, 'and I just feel terrible about it. I know this does not sound very professional of me, but perhaps she just missed you so much that she died of a broken heart.'

An Imaginary Continuity

Intimate connection with the natural world may be our most primal and important cultural heritage, but it's also true that today we can only dimly remember it. We have to work hard to stay in touch, through our pets to a degree, but mostly with words. We enshrine the natural world with popular idioms such as 'Mother Earth' and 'Mother Nature', although 'Stepmother Earth' and 'Stepmother Nature' might be more accurate at the end of this millennium. National parks, private gardens, municipal zoos, household pets and natural history films are the closest most of us can get to Mother Earth nowadays. In my opinion, there is not much point in tearing our hair out over the loss of our sense of intimate belonging in the wide world of flora and fauna. That's just the way it is, and there will be no 'going back', not that I can see. This particular horse is definitely out of the stable, and sprinting away as fast as it can.

Is our anthropomorphic instinct one way in which we try to stay in touch with the natural world? Many commentators believe so. We see the golden eagle riding the thermals above a beautiful valley and imagine ourselves in that situation, and become envious. What sheer fun that must be, what joy, we believe, even though we have absolutely no idea if this is the eagle's own emotion. Or we see a dog cowering in the corner after a severe reprimand from his or her owner and we feel sorry for the poor thing — or a little guilty, perhaps, if we are the owner. It seems that we cannot help projecting our attitudes, emotions, and ideas of personality and rationality onto – or into – our household pets and many other animals as well.

Anthropomorphism is also a reflection of our 'psychic structure', to quote Paul Shepard's book *The Others: How Animals Made Us Human*. Before his death in 1996, Shepard was a professor emeritus of human ecology at Pitzer College. He suggests, in this wonderful book, that animal stories and anthropomorphising of every sort set up an 'imaginary continuity between animals' lives and our own [providing] a lifelong shield against alienation . . . Stories with animals are older than history and better than philosophy.'

That is very well said. Stories with animals also make the best cartoons. Many of our most creative cartoonists would be hard pressed for subject matter if they couldn't depict talking, thinking, emoting animals. A quick survey of Greg Larson's 1998 'Far Side' desk calendar finds 16 such jests in the month of August alone. That's 16 out of 31 – and all of them terrifically clever. Take the scene on Noah's ark with all the creatures gathered around what we assume to be a fresh carcass. (All we see of the deceased is its extended hooves, sticking straight up in the air.) Hands on hips, and glaring in the direction of the lions, Noah declares sternly, 'Well, so much for the unicorn, but, from now on, all carnivores will be confined to "C" deck.' The lions look knowingly at each other: classic anthropomorphism with a wicked twist, because we presume that one thing lions do *not* feel is guilt about their carnivorous nature.

Or would Charles Darwin disagree? In his defence of animal consciousness, the great evolutionist was a committed, inveterate anthropomorphiser. *The Descent of Man, and Selection in Relation to Sex*, published in 1871 (a dozen years after *The Origin of Species*), is full of passages such as this one: 'We have seen that the senses and intuitions, the various emotions and faculties, such as love, memory, attention and curiosity, imitation, reason etc., of which man boasts, may be found in an incipient, or even sometimes in a well-developed condition, in the lower animals.'

Getting down to details, Darwin wrote, '. . . a dog carrying a basket for his master exhibits in a high degree self-complacency or pride. There can, I think, be no doubt that a dog feels shame . . . and something very like modesty when begging too often for food. A great dog scorns the snarling of a little dog, and this may be called magnanimity.'

Perhaps the most famous such Darwinian passage among biologists is the reference in *Descent* to 'ants chasing and pretending to bite each other, like so many puppies'. One finds that notorious excerpt in quite a few commentaries! To the scientists who followed in Darwin's footsteps and defended his controversial theory, such passages from *Descent* and others from *The Expression of the Emotions in Man and Animals* were acutely embarrassing. Since these scientists don't like to criticise Darwin, for fear of casting doubt on his main theory, to this day they have tended to focus the blame for this kind of blatant anthropomorphising on his protégé, G.J. Romanes, who wrote in the same vein in his book *Animal Intelligence* (published in 1882). Romanes's narrative is chock full of stories about jealous fish and dogs who enjoy a knee-slapping joke.

The Quandary

In her book *Pack of Two*, Caroline Knapp describes her deep love for her dog Lucille and her struggle to balance this love against the well-known human tendency to anthropomorphise. Knapp describes how, at least once a day, Lucille:

> . . . pokes her nose toward my hand. This is the kind of moment that can make me feel in tune with her, that seems to speak to a level of connection between us that I cherish: she is asking me to give her my hand, and when I comply she will sit down and spend several minutes licking it . . . she sometimes places a paw on my wrist to steady my hand, a gesture that feels delicate and tender and full of affection to me. But is it? I'd like to think that this is about love, that Lucille

is deliberately seeking me out for a kind of canine kissfest, but it's equally possible that she's indulging one of her other affinities: for the taste of hand lotion. So which interpretation is right? I choose to believe that the hand-licking is about both moisturizer and me (they're my hands, after all), but I am also aware that an element of choice is at work here, that this is my read, based at least to some extent on my own investment in the idea of loving communion with my dog.

Just so, says Celia Heyes, a comparative psychologist at University College, London. Heyes is not a full-blown Cartesian on the question of animal consciousness, but she is an agnostic. We don't know whether animals are conscious, she argues. We don't know whether dogs truly 'love' us, but we do know that seeing ourselves in them seems to be our pre-set 'default' mode of thinking.

In his book *Bonobo: The Forgotten Ape*, Frans de Waal, a primate researcher, writes:

> When the lively, penetrating eyes lock with ours and challenge us to reveal who we are, we know right away that we are not looking at a 'mere' animal, but at a creature of considerable intellect with a secure sense of its place in the world. We are meeting a member of the same tailless, flat-chested, long-armed primate family to which we ourselves and only a handful of other species belong. We feel the age-old connection before we can stop to think, as people are wont to do, how different we are.

In short, our tendency to anthropomorphise is the 'default' mode only because it reflects a deep and vital truth: how else can we view the world, except by listening carefully to our own knowledge and experience, and by reacting within the categories our brains allow? When we look into the eyes of a dog, we *do* feel the heart-to-heart connection. To deny this fact is silly and, in the end, fruitless. When we look into the eyes of a bonobo or a chimpanzee, we *do* see 'a creature of considerable intellect with a secure sense of its place in the world'. René Descartes himself had a pet dog, Monsieur Grat, and I'm willing to bet that he treated him kindly, even while insisting that the animal was merely a machine.

In the early years of the *Nature* television series, we bent over backwards not to be anthropomorphic in the narration scripts for our programmes. We wanted our films to be scientifically accurate and to be taken seriously. We still do, of course, but the more we learn about the animal mind, the more inclined we are to give the subjects of our films greater credit for the intelligent behaviour they exhibit in front of our cameras.

Celia Heyes argues that scientists and others who would know the truth must watch out for the distorting anthropomorphic bias. Fair enough: we should be careful when we're doing science. But when, in the name of science, we insist on viewing the wild world with absolutely no sense of 'connection', the result is just as unsatisfactory and false as the most far-fetched nineteenth-century tale about the dog who enjoyed a joke.

chapter 2

BEHAVIOURISM AND BACKLASH

The Basic Agenda

If I've read about Clever Hans once, I've read about him fifty times. It must be illegal to omit this famous horse from any consideration of our ideas about the animal mind, and I don't want to pay the fine. Hans's owner was Wilhelm von Osten, a Prussian aristocrat who came into a sizeable fortune around the turn of the century and immediately set out to prove his prowess as a teacher by instructing his horse in simple arithmetic. And he succeeded in doing so, at least to his own and to many observers' satisfaction. If von Osten wrote down on a card '2 + 4', Hans tapped his forefoot six times; if he wrote '3 × 3', Hans tapped nine times. Even when another individual was in charge of the examination, Hans was usually correct in his computations. He became a sensation, a regular Secretariat in regards to fame.

In 1904, Hans even passed a set of what were designed to be rigorous tests by the director of the Berlin Psychological Institute, one Carl Stumpf. The panel of judges included a circus trainer, a zoologist, a veterinarian and a politician. Stumped by the horse, Stumpf then turned over the case to his pupil Oskar Pfungst, who ruined all the fun by demonstrating that Hans was not doing arithmetic at all. Instead, he was responding to the very subtle visual cues of his inquisitors.

Now, von Osten was not a con man. Apparently he was truly unaware that he had been subtly cueing his horse, just as the other testers had been unaware of their own inadvertent cues. But Pfungst proved that von Osten and the others had indeed been tipping the animal off. Pfungst did so quite simply, by directing the examiner to pose problems written on cards that the examiner could not see. In this case, poor Hans was lost and pawed the ground at random. Pfungst went on to demonstrate that Hans could respond to cues as slight as the raising of an eyebrow or the flaring of the nostril. He was indeed a clever horse, just not in the way his owner wanted to believe.

The embarrassment of the Clever Hans fiasco hung over animal studies and comparative psychology for decades, and it still does, to some extent. Indeed, the episode amply demonstrated the dangers of the anthropomorphic bias and the necessity for rigorous procedures in the scientific evaluation of animal behaviour. Case closed. It is not much of an exaggeration to say that many researchers, especially those based in America, diligently de-anthropomorphised their approach in the early part of the twentieth century – even if this took them all the way back to Cartesian ideas about animals as nothing more than four-legged machines.

From this perspective, there was no place for 'anecdotalism', either – with anecdotalism defined as almost any observation obtained through naturalists' old-fashioned, patient observation in the field. Like anthropomorphism, this allegation of anecdotalism has been quite important in animal studies throughout this century; I want to present a couple of field-work episodes from quite different types of sources to illustrate the problem.

The first comes from *Elephant Memories*, Cynthia Moss's compelling book about her 13 years living among and studying the elephants in Amboseli National Park in Kenya. In 1980, twins were born to the Amboseli elephant named Estella. (Even the naming of animals in the field, as opposed to 'objective' numbering, is looked on with suspicion by some scientists.) One of these twins, Equinox, was male; Eclipse was female. Twins are a rarity among the elephants, and surviving twins even more rare, presumably because the mother can provide only enough milk for one. So, Cynthia Moss thought, this will be quite interesting. Sex differences should come through loud and clear in this 'experiment'. Indeed, the first time Moss observed Estella nursing the two calves, the male Equinox bullied Eclipse away from the teat on one side, then on the other. Moss writes, 'It was always the male, who was slightly bigger, who initiated the aggression, and he was usually successful at disrupting the female's suckling.'

Within a week, our veteran observer Moss was beginning to doubt that the female would survive, but then one day she watched as the two young elephants ran around and butted and played with each other for ten minutes. Finally the young male got tired and lay down to rest. Immediately Eclipse ran to the teat on the far side of their mother Estella, out of sight of her brother (who was asleep anyway), where she suckled in peace. So this was how the calf had been obtaining her milk ration! Both calves survived and thrived.

Now consider this bizarre episode from Elizabeth Marshall Thomas's best-selling book *The Hidden Life of Dogs*, which is unabashedly, unapologetically anthropomorphic and anecdotal. But how could our relationship with dogs be otherwise, Thomas pleads, after thousands of years of co-dependency (my term, not hers!) between the two species? The story concerns her husband and one of the couple's dogs. One blistering hot afternoon in Boulder, Colorado, where they lived at the time, Thomas's husband bought an ice cream cone. When he started licking, he noticed the dog watching with close attention. Sympathetic, he offered the cone to the dog, fully expecting to lose the whole thing with this altruistic gesture. To his surprise, the dog licked daintily and stopped. Man and dog exchanged polite licks a second time, then a third, until they were down to the cone itself (which some of us feel is the best part). Now Thomas's husband took a small bite and offered the stump to the dog, again expecting to lose the rest of it. Instead, the animal astonished everyone again by taking only a small bite, just as his master had. Man and dog then exchanged several more considerate bites.

To Thomas, this behaviour was by no means astonishing. She explains:

For eight years, my husband and this dog have built a

relationship of trust and mutual obligation . . . when two dogs share food, they eat simultaneously while respecting each other's feeding space, which is a little imaginary circle around the other's mouth. But the idea of taking alternate bites is totally human. Even so, the dog fathomed it, and without ever having seen it done. Who ate the tip of the cone? My husband ate it. The dog let him have the last turn.

Do these two stories represent irrelevant 'anecdote' or invaluable observation? The anti-anthropomorphising camp would reject both out of hand as nothing more than whimsical anecdote – wishful thinking, folk psychology, Clever Hans – all over again. Their view is encapsulated in the standing joke 'my data, your observations, their anecdotes'. And it is true that such evidence must be carefully sifted, but it is also true that the rare behaviour serendipitously observed might well be exactly what is most valuable for our understanding of any animal.

Cynthia Moss is an elephant expert of many years' experience. I think her story of Estella and the twins might well demonstrate shrewd thinking by the little female. And here's what Cynthia has to say about the practice of culling elephants:

> They cull whole families except for the youngest calves and then capture those for sale. And everyone said, 'Oh, that was okay,' because the whole family was killed and no other elephants knew about it. But now we find through our field studies that elephants can hear over long distances – these infrasonic sounds – so they can hear the screams – the death screams of those elephants – maybe from two kilometers away. And then those elephants that remain alive are terrorized and every time they hear the helicopters go up, they're afraid. They live lives of terror, which is something that we cannot accept.

If we did set aside all such testimony from the field, what could we use for evidence in its stead? At the turn of the century, the answer of the comparative psychologists was . . . laboratory testing. In Russia, Ivan Pavlov conducted his famous experiments demonstrating conditioned reflexes in salivating dogs. In America, the psychologist and educator Edward Lee Thorndike contrived a classic experiment involving a puzzle box from which famished cats were challenged to escape by finding the correct sequence of release mechanisms. The only way the animal could accomplish this the first time was to happen across the right sequence by pure luck. (The same would have been true for a human being.) A plate of seafood was the cat's reward. With subsequent trials, the time required for every cat in the experiment to effect its release got progressively shorter, until eventually the animals could escape almost immediately.

Thorndike considered two explanations. This was either a straight-forward case of stimulus-response learning, the product of reinforcement; or the cat was rationally thinking its way through the problem, using prior results as a guide. One factor that inclined Thorndike towards the stimulus-response explanation was that the successfully trained cat would follow the release sequence even if one side of the puzzle cage had been removed and the cat could simply have stepped out and started eating. Given this fact, how could the cat be reasoning? Thorndike decided it couldn't be, and he derived what became known as the 'law of effect', which stated that behaviour changes because of its direct, immediate consequences and for no other reason. (Later in his career he turned to creating intelligence and learning tests for children.)

In 1913, J.B. Watson codified many of the tenets of Pavlov, Thorndike and other laboratory researchers and thereby founded the field within psychology that became known as behaviourism. (Watson then had a different kind of 'after-life' from Thorndike. When a sex scandal ruined his career at Johns Hopkins University, he went to work for the J. Walter Thompson advertising agency on Madison Avenue. In the annals of advertising, he is often credited with inventing the 'brand name'.)

Although Watson was the official founder of the behaviourist school, these ideas gained public, as opposed to just academic, attention through the notorious pronouncements and investigations of Burrhus Frederic Skinner, who is practically unknown by that name but world-famous by his initials: B.F. Skinner. Operating from his powerful position at Harvard University, Skinner succeeded in replacing the overly enthusiastic anthropomorphising of nineteenth-century naturalists such as Darwin and Romanes with the equally unrewarding, reductive philosophy of behaviourism. This approach held sway in comparative psychology for much of this century, and is still highly influential. Today, the field of cognitive ethology – the study of animal cognition, minds and consciousness – is, in many respects, a backlash against behaviourism.

Like just about every other school of thought, this one evolved and split into various subtly differentiated factions, but the behaviourists were united in their belief that, as far as scientific investigation of animals goes, all that matters is observable behaviour, because only such behaviour can be studied with reliability, objectivity and repeatability – three benchmarks of the scientific method. Everything else is without 'cash value'. We might believe we can intuit thoughts and feelings in animals, but where's the proof? We don't have any. Furthermore, behaviourists argue, we never will. We can no more scientifically study animal consciousness than we can poetry. The 'common sense' of anthropomorphism can be no guide because in previous eras common sense has told us that the sun revolves around the earth and that space and time are constants – how could they be anything else?! But those apparently self-evident truths have been proved incorrect. Thus there is no place in science for 'common sense'.

I pause to look over at my three dogs and one cat and I wonder, 'What are these creatures thinking?' I ask the same question about all the animals I have observed in the wild. Behaviourists answer, 'They're not "thinking" anything at all, and they're not "feeling" anything, either. They're *behaving*, often in apparently sophis-ticated and provocative ways, granted, but still just behaving.' Or, as Skinner wrote, in *About Behaviorism*, 'Some [mental states] can be "translated into behavior," others discarded as unnecessary or meaningless.'

Even what we laypeople think of as pure native instinct — breathing, to use an extreme example — was viewed by behaviourists as stimulus-response learning accomplished very quickly, and at a very early age! Behaviourists didn't even try to 'understand' their subject. That enterprise, too, was beside the point and loaded with anthropomorphic pitfalls. The best that could be hoped for was accurate, complete description and, perhaps, control. (Skinner frankly acknowledged that his big idea had been a philosophy before it became a way of doing science. And, as a philosophy, behaviourism had profound implications, as implied by the title of his most inflammatory book: *Beyond Freedom and Dignity*.)

The Interpretation of Dreams, written by Sigmund Freud in 1900, was translated into English in 1913, the year in which J.B. Watson codified behaviourism. Needless to say, behaviourists were not impressed by Freudianism, which they considered totally unscientific. When Freudians in turn criticised behaviourism for being unable to deal with the unconscious, Skinner pointed out — with great delight, I imagine — that in fact his approach deals with almost nothing *but* the unconscious. After all, we're unconscious 24 hours a day of what goes on in our brains. Behaviourists just never used the term in its Freudian sense.

They always took great care with language, which was redefined as 'verbal behaviour'. (Skinner even wrote a book with that title.) Love was 'attachment formation'. The self was a 'repertoire of behaviour'. They wouldn't use a phrase like 'the dog threatened the cat' because the term 'threaten' anthropomorphically implies some kind of mental state (some intention or purpose) on the part of the animal. They might not even have said the dog 'raised its right leg', because this innocuous verb implies the intention to do so. It would be safer to put the phrase in the passive voice: 'the right leg is raised' or 'the right leg goes up'.

As the joke goes, no behaviourist ever changed his or her mind! They became notorious for believing that we do not believe and for thinking that we do not think. It seemed that behaviourism had contracted a bad case of 'physics envy', a widespread malady in twentieth-century science, and academic disciplines in general. Many academics (not just psychologists, by any means) labour in apparent jealousy of the 'hard and fast truths' discovered by physicists and chemists and mathematicians. Never mind that these truths sometimes turn out to be false; at least there is the prospect of some solid answers. The incredible discoveries of what we know as the 'hard sciences', and the technologies that quickly followed these discoveries, established such disciplines as the only purveyors of Truth with a capital 'T', and comparative psychologists wanted in on the action.

Eventually, behaviourism melded with the equally notorious socio-biology (as formulated by E.O. Wilson, the famous authority on insects), to yield a more or less unified, ultra-materialist approach to behaviour, consciousness and mind. This viewpoint hypothesises that all aspects of behaviour and culture – human and animal – can be accounted for by a combination of genetic inheritance, handed down through uncounted generations by way of natural selection (the focus of sociobiology), and conditioned behaviour 'learned' by our brains by way of reinforcement (the focus of behaviourism). The essence of sociobiology, which was widely derided when it first came to public attention in 1975, is now somewhat back in vogue under the new label 'evolutionary biology', without much controversy and with rave reviews for Wilson's latest book, *Consilience*. (It's easy enough to challenge this viewpoint, and I do. But it is much more subtle than it might appear at first glance. I am simply not ready to admit that the brain is nothing more than the world's best computer; there's too much magic in there.)

A Byte is Not a Bite

So we have behaviourism, aided and abetted by physics envy, operating throughout this century as the single most important influence in comparative psychology and the cognitive sciences. Since the 1960s, a second major influence has been computer science and ideas about 'information processing'. The most radical cognitive scientists have concluded that 'mind' is nothing more than our experience of the brain, in some way. 'The mind is what the brain does,' according to Steven Pinker (linguist and author of *How the Mind Works*). The brain is nothing more than an organ for 'information processing' – a fantastically sophisticated, carbon-based computer. These theories can get unbelievably complicated. But, like behaviourism, they still constitute a profoundly reductive, mechanistic approach to the question of consciousness (including our own).

It is undoubtedly true that a great deal of animal and human behaviour amounts to unconscious information processing of some sort. After a complex task is initially learned, it can be accomplished for long periods of time without any sense of conscious awareness. In fact, conscious effort during such tasks just causes problems. The pianist, for example, who laboriously learns a concerto cannot then play that music in a concert with the same 'conscious' attention it took to memorise it. If he or she tried to think consciously about every note – disaster! With these difficult skills, the whole point is to transform conscious learning into muscle memory and then to 'let it flow'.

A more common example of unconscious information processing – the one quoted in almost every book on the subject – is the task of driving a car, which is quite complex, when you stop to think about it. The point is that, after we have put the key in the ignition, we *don't* stop to think about it. We just drive, and we realise we have been totally inattentive – engaged in conversation with a friend or listening to the radio – until a sudden event, such as a car pulling out into our lane, draws our attention back to the task in hand. Then we become aware that we are, in fact, driving. And when I think carefully about talking to someone when I am driving a car, I realise that I don't consciously 'choose' every word I say. In fact, I don't consciously choose many words at all. Suddenly, I've just said them, and I suspect that this is also the case with whoever I'm talking to.

Likewise, as I sit at my computer writing these pages, 'I' know I consciously chose the subject, but the specific words in this sentence are suddenly just here. 'I' haven't consciously sorted through the confines of my brain to select them, one by one. So there is a good deal of conscious as well as unconscious information processing going on. Still, it is an incredible leap from these straightforward, less-than-profound insights to the 'nothing but' conclusion. This is the part of the argument that bothers me most.

How would scientists and philosophers who hold this 'nothing but' position go about proving it? They might start with the famous test for artificial intelligence devised by the brilliant mathematician Alan Turing. Turing proposed that if a computer behind a screen could fool an interlocutor in an extended conversation, then that computer must be considered 'intelligent' in the same way that we are intelligent. No computer has passed the Turing test, but I wonder whether it's a legitimate challenge in the first place. Let's consider an analogous endeavour in the biological sciences: the creation of a robot that could perfectly mimic the behaviour of a laboratory mouse.

Suppose for a moment that a team of computer whizzkids does succeed in creating a little robot that perfectly mimics the behaviour of the lab mouse in all observable respects. I've seen television footage of the latest robots that roll around on wheels, guided by laser eyes, and successfully avoid all the bigger obstacles. I cannot honestly envision the upgraded, even perfected version that could match the mouse in all respects, but let's suppose it is built. And let's suppose the behaviour is identical. This would be an extraordinary technical feat, but would anyone seriously argue that there is *no difference at all* between this incredibly sophisticated little machine and this relatively 'dumb' little animal? I really doubt it.

Remember that the incredibly cruel experiments on animals in the seventeenth century backfired, in a way, because they mainly demonstrated the almost identical physiology of all vertebrates. (As

Voltaire exclaimed, 'Answer me, mechanist, has Nature arranged all the springs of feeling in this animal to the end that he might not feel?') Any twenty-first century achievement in the field of artificial intelligence will only serve to make a related distinction. Even if the lab mouse is defined as a carbon-based information processor – even if the brain is, as artificial intelligence guru Marvin Minsky once famously said, 'a computer made of meat' – there remains an undeniable difference between the two test subjects. One is a mobile computer plugged into the wall; one is a mouse with at least minimal sentience. It's that simple. The behaviour might be the same, but the means of achieving it are different. No theory can make it otherwise, can it?

The Backlash

The influential linguist Noam Chomsky once said that human ignorance can be divided into two categories: problems and mysteries. Problems might be solved one day, while mysteries are beyond solution. Until fairly recently, the cognitive sciences seemed to have got stuck investigating the mystery of consciousness. In 1973, the philosopher Thomas Nagel published what became a famous paper on the subject entitled, 'What is it like to be a bat?' When trying to answer that question, Nagel said he found himself 'restricted to the resources of my own mind, and those resources are inadequate to the task'. He answered, in effect, that we'll never know what it's like to be a bat.

Twenty-five years later, some researchers and thinkers in the cognitive sciences argue that we may end up surprised at how much we can learn about what it's like to be a bat or any other creature. No one believes that we'll ever answer every question we might have about animal or human consciousness, but there's little doubt that scientists working in a variety of fields are moving some of these questions out of the category of mysteries and into the category of problems.

Two thousand years ago, astronomy and medicine, for example, were only marginally scientific; our ancestors didn't have the necessary tools. Three hundred years ago, Sir Isaac Newton discovered the law of gravitation and (along with Gottfried Wilhelm Leibniz) developed the calculus – and then he turned his attention to the wonderful world of alchemy because there was no modern science of chemistry. A hundred years ago, we had no way to study the brain and corresponding issues of consciousness. Today, we can. Yesterday's EEGs were crude instruments compared with today's CAT scans and PET scans and ERPs (event related potentials). Tomorrow? There will be something new – that we can be sure of. Later we shall consider the fascinating research in this area. But, at this point, suffice to say that we know that certain ERP activity in the human brain is associated with very complex processing of data, and we know that ERP activity in other mammals is almost identical to ours in the same circumstances. This proves nothing, but it certainly raises questions for further research; as does the fact that psychoactive drugs have exactly the same effect on cats and dogs (and other animals, presumably, although pets are the most extensively treated in this way) as they have on humans.

In philosophy there is a bizarre notion labelled 'philosophical solipsism'. It has many variants, naturally, but the main one is the idea that I am definitely a conscious human being but you might be a robot, and I can never prove otherwise, because I cannot get inside your head. This notion is taken seriously by few philosophers because of all the inferential evidence to the contrary, but many of the objections to intelligence and consciousness in animals are rooted in an analogous 'species solipsism': we humans are conscious, but we will never know about you dogs. But this solipsism is no more valid than the other kind. The operative scientific phrase is 'inference to the best explanation'.

I was surprised to learn from Donald Griffin (retired Professor Emeritus of Zoology at Rockefeller and Harvard Universities) that quarks, the building blocks of all matter, according to the best theories, are actually inferences from the data. Molecules and atoms we can see in electron microscopes, but we cannot see quarks. We infer their existence. Likewise, very little of the evidence supporting the evolution of species has been observed with our own eyes and instruments. Evolution is an inference from the record. Yet many scientists refuse to consider the reality of mental states in animals on the grounds that such states cannot be observed! A glaring inconsistency, as Donald Griffin points out.

No matter how sophisticated the behaviour, behaviourists can always simply assert that 'it's learned'. In fact, that's exactly what radical behaviourism does assert. In order to construct the theoretical framework for such a narrow approach, they are compelled to diminish our sense not only of animals but of ourselves. Furthermore, their reductionism is not even good science. It is true that no one can ever prove that animals have consciousness (just as I cannot prove you're not a robot), but to declare today that animals absolutely cannot think or feel is no different from medieval theologians testifying that the planet earth cannot possibly orbit the sun. It is just as unscientific as the anecdotal, anthropomorphic assertion that when his owner leaves him alone in the house for ten hours, Fido gets 'depressed'.

Now the tide is turning. Over the past 25 years, an ever-increasing number of ethologists have balked at such arbitrary reductionism. The growing body of evidence we will survey in the following chapters clearly demonstrates that there is more to animal behaviour than just instinct and conditioned learning. There is, at the very least, extensive memory and cognition. Even Pavlov's notorious dog experiments, interpreted correctly, showed more than knee-jerk salivation. But, operating in that early-behaviourist, radically anti-anthropomorphic environment, the Russian could not or did not make the connections. (Nor did he stop to ponder the fact that his experiments drove some of his lab dogs crazy.)

Restricted as they were to artificial settings, many of the laboratory experiments of the comparative psychologists asked narrowly restricted questions and received in return narrowly restricted answers. To thoroughly investigate animal intelligence, we need to observe and understand them in the environment in which they evolved, where their behaviour will make the most sense. Their intelligence is quite narrowly focused. As we shall see, creatures capable of astonishing mental feats in one area can accomplish little in another area that does not seem much more difficult. (If you think this point does not apply to us humans, consider the two skills, language and arithmetic. Language should be extraordinarily difficult – but we pick it up automatically. Arithmetic should be easy by comparison – but months and years of focused effort are required before we master the basic skills.)

Of course, many biologists and naturalists have always understood the importance of field work, even in the heyday of behaviourism. In Europe, especially, there has been a long tradition of field-based naturalism. The validity of field studies is much greater in this century, with much stricter criteria and with much more care taken to account for the biases of the observer. Wolfgang Kohler carried out some groundbreaking 1920s experiments on the tool-using capabilities of captive chimpanzees, and concluded that animals employ insight and purpose in their behaviour and lives. (More importantly, Jane Goodall later proved that chimps living in the wild also used tools.) Konrad Lorenz, the famous researcher whose work with greylag geese nailed down the concept of 'imprinting', wrote openly of emotions in animals in his book *Man Meets Dog* and elsewhere. In Japan, studies in ethology have always been fundamentally anthropomorphic, because Japanese culture in general has never accepted the Western cult of rationality.

The concept of the *umwelt*, used by the nineteenth-century physiologist Jakob von Uexkull to mean the specific sensory world of any given animal, has been deemed of great importance by men and women who actually observe animals in the wild. What is the *umwelt* of Thomas Nagel's bat? We can't be certain, but we do know that it is very different from our own. A dog's *umwelt*, heavily reliant as it is on smell as opposed to eyesight, must be drastically different from ours. How can laboratory experiments ever get to the heart of these issues?

Sue Savage-Rumbaugh writes, in *Kanzi: The Ape at the Brink of the Human Mind*:

> Would the error of sometimes erroneously attributing capacities that did not exist be greater than that of never discovering any emotional or intellectual capacities that were continuous with our own? I think not. At the very least, if one scientist made a mistake and attributed some capacity to an animal that was far beyond its true ability, another scientist would come along and correct this mistake . . . [The Clever Hans episode is a perfect example.] As the situation currently stands, we don't even have the right to make the mistake.

Kanzi was published in 1994, and perhaps Savage-Rumbaugh exaggerates. She has a lot of allies, including the Tufts philosopher Daniel Dennett, whom I will mention often because he's one of the most cogent thinkers about mind and consciousness. Dennett disputes the behaviourist approach. He argues the importance of questions such as 'Why does the fiddler crab have this one enlarged claw? What does he intend to accomplish with it?' Behaviourists recoil in horror, but in asking such questions we're not necessarily saying that the crab in fact does know why. We address such questions in order to learn answers from evolution. Even the simplest bacterium has its reasons, Dennett writes, with his usual wit. It isn't aware of them, true, but for us to ignore these reasons in our investigations is folly.

This is *pragmatic* anthropomorphism, an important idea which is nicely illustrated for us by one Jerry Garcia – no, not the deceased founder of the Grateful Dead, but the famously inventive experimental psychologist in California. Garcia wrote, in *American Psychologist,* that he uses anthropomorphism to predict animal behaviour 'because this works better than most learning theories. I could rationalize this heresy by pointing to our common neurosensory systems or to convergent evolutionary forces. But, in truth, I merely put myself in the animal's place.'

A Parting Shot

I conclude this brief discussion of the behaviourist philosophy and the backlash against it with a remark by Dr Nicholas Dodman, Professor of Behavioral Pharmacology and Director of the Behavior Clinic at the Tufts University School of Veterinary Medicine. (I think of him as the pet psychiatrist.) 'Look,' Dodman said to me in a Thai restaurant near Tufts, 'if it walks like a duck and talks like a duck, we have to investigate the possibility that it might be a duck.'

chapter 3

THE COGNITIVE REVOLUTION

A Fresh Start

The last two decades have witnessed nothing less than a 'cognitive revolution'. The blasphemous idea that even the lowly duck might have mental states is now in play, and the scientist most responsible for the increasing acceptance of cognitive ethology as a valid field of study is Donald Griffin. He argues that students of animal behaviour have been 'haunted' by the spectre of Clever Hans errors for long enough. I've mentioned pragmatic anthropomorphism; Griffin's preferred term for the new approach he advocates is 'critical anthropomorphism'. His books *Animal Thinking* and *Animal Minds* were the groundbreaking texts in this field, and a good deal of my understanding of the major issues in cognitive ethology comes from these two volumes, from one long conversation with him in his Harvard office, and from my reading from his forbiddingly enormous, but authoritative, bibliography.

Griffin's first area of study was the mysterious bat. Before his research, our knowledge of bat orientation and navigation was based mainly on the eighteenth-century experiments of Lazzaro Spallanzani, an Italian biologist who was able to establish the important fact that bats primarily use their ears for navigation. The proof was pretty simple, if cruel: subject bats that had had their eyes removed flew around as well as normal bats; other bats with functioning eyes but with their ears plugged ran into problems – or *flew* into problems, to be more exact. However, this conclusion was not widely accepted because no human could hear what the bats were hearing. It was Griffin who proved that these remarkable mammals were hearing ultrasound – their own ultrasound.

In his research, which began while he was still an undergraduate at Harvard, Griffin got off to a good start by simply pointing a new-fangled ultrasound detector at a bat hanging in its cage. The meter went off the chart. So these bats were emitting ultrasound waves. Further work in the lab and outdoors proved that they were emitting the waves from their mouths (because bats trying to navigate with their mouths taped shut had trouble). Griffin proved that the rate of ultrasound emission by the bats increased dramatically when they flew within a metre or so of an object that could be as tiny as a strand of wire. Therefore an insect, even one on the wing, is a huge target, a veritable piece of cake, for a hungry bat. Griffin named the process by which bats navigate 'echolocation'; his book *Listening in the Dark* was published in 1958.

Providentially, perhaps, Griffin encountered Thomas Nagel just as the philosopher was mulling over the ideas that would appear in the seminal paper 'What is it like to be a bat?' Griffin knew more about this subject than anyone on earth, but it was a long conversation with Nagel that first got him interested in the broader questions of consciousness. While Nagel eventually concluded that the *umwelt* of the bat is so different that we'll never know what its subjective experience is like, Griffin believes that we may be able to learn a lot more than we realise.

At the age of 83, Griffin doesn't give many interviews these days, but he agreed to see me in the small office he keeps in the old Anthropology Museum at Harvard. 'It's a creaky old elevator,' he had warned over the telephone. 'You push the button for the fourth floor and then for some unknown reason you have to push the unmarked button beside the fourth-floor button. Now don't forget to push both.' This sounded like one of Thorndike's puzzle boxes for testing cats. What would happen if I did forget, which was entirely possible? But I negotiated the terrain and Griffin greeted me at the doorway to his office. He is tall and distinguished and soft-spoken, hale and hearty and much younger-looking than his age – not exactly how I had visualised such an iconoclastic scholar.

He said almost immediately, 'I think everyone would agree that the behaviourist movement started by John Watson and continued by B.F. Skinner here at Harvard was very influential in discouraging, let's say, interest in subjective animal mentality and animal consciousness – up until very recently. Indeed, there are still some strong advocates of the behaviourist point of view.'

Because of the explosion of scholarly interest in the subject of animal minds, it's easy to forget that the cognitive revolution faces serious challenges from within academia. In fact, Griffin even suspects his own view may still only be held by a minority. 'There have been a number of people who have been, if not suspicious, at least very cautious,' Griffin continues. 'One of the leading ones is a Canadian psychologist named Hank Davis. He honoured me by likening my book to Salman Rushdie's *Satanic Verses*.' Griffin gets a chuckle out of that somewhat hyperbolic denunciation. Another 'debunker' I read argued strenuously that the best label for the ideas of the cognitive ethologists isn't 'pragmatic' or 'critical' anthropomorphism, but 'anecdotal cognitivism'. In the trade, that's a harsh put-down. In a review of Griffin's *Animal Thinking* in *American Scientist*, E.A. Wasserman wrote, 'No statement concerning consciousness in animals is open to verification and experimentation. Isn't it time we set aside such tantalizing, but unanswerable, questions and direct our energies to more productive pursuits?'

Ironically, it was Wasserman and colleagues who were responsible for some of the first tests demonstrating that 'the conceptual abilities of pigeons are more advanced than hitherto suspected'. These famous tests demonstrated the powerful category-creating capacity of the lowly, 'bird-brained' pigeon. In some tests, the birds outperform college students, time and again. We will look at these experiments in detail in Chapter 4. Yet the Wasserman paper presenting these findings was entitled 'Conceptual behavior in pigeons'.

Behaviour. That's the giveaway. As Griffin comments:

Animals may behave as though they utilized simple concepts, but behaviourists are constrained to ignore or deny
the possibility that they might consciously think about the
categories or concepts. To summarize briefly, the study of
animal cognition is now scientifically respectable, but not
animal consciousness. In other words, cognition in the sense
of information processing in nervous systems, that's okay.
But there is a great reluctance to take the further step of
saying, well, maybe if there's cognition there is also conscious
thinking.

Definitions

Cognition? Consciousness? The time has come for some definitions.
The famous Berkeley psychologist David Krech said, 'There is no
phenomenon, however complex, which when examined carefully will
not turn out to be even more complex.' This is definitely true in the
cognitive sciences. Almost every philosopher since Plato and Aristotle
has wrestled with the problem of consciousness and mind in humans,
and many have also pondered the nature of animal minds. In this
century, many of the discussions seem to have got bogged down in
questions of semantics. (The same is true of other academic disciplines
as well. Before modern philosophers can discuss 'love' or 'virtue' or
'good' or 'God', they have to define these words, and the task turns
out to be more difficult than Plato and Aristotle ever imagined. Or
so today's deep thinkers have concluded.)

All in all, splitting hairs is child's play compared with drawing
meaningful distinctions between 'sentience' and 'cognition', between
'cognition' and 'thinking', between 'thinking' and 'consciousness',
between 'consciousness' and 'self-consciousness'. In their fine book
Species of Mind: The Philosophy and Biology of Cognitive Ethology, Colin
Allen and Marc Bekoff refer teasingly (Allen is a professor of
philosophy) to 'the cold touch of technical philosophy'. I know

the feeling. Sometimes I found all the academic rigour and necessary scepticism frustrating, but it was also exciting and challenging. For the behaviourists, the confusion is a godsend. They propose something called the 'definition objection', which argues that if we can't come up with a strict definition of 'consciousness' and other terms, there's nothing to study.

Griffin is ready for this one. He pointed out to me that one of the dirty little secrets of science is that there's no really clear and agreed-upon definition of 'gene', either. It is not just a specified sequence of nucleotides on a chromosome, but something much more elusive. Still, thousands of scientists the world over are studying genes, with remarkable results. What's more, he added, 'You can't define motivation or learning either, if you want to get increasingly rigorous. All of these are somewhat vague and general concepts. So I am not discouraged by the fact that it's difficult to define consciousness. I think if we knew more about it, it might come easier.'

For the definitions that will be 'operative' in this book I turn – logically enough, I think – to *Cognitive Ethology*, the volume of essays collected in honour of Donald Griffin in 1991. Here 'cognition' is defined as 'processes by which the sensory input is transformed, reduced, elaborated . . .' Closely related to cognition is 'thinking', defined as 'attending to the animal's internal mental images or representations'.

As laypeople, we might consider cognition and thinking to be more or less synonymous, but the term 'cognition' is preferred by those researchers who intend to deny any kind of subjectivity or consciousness for a given animal. They shy away from the word 'thinking', which seems to imply just a hint of those *verboten* qualities.

Griffin explained to me that almost everyone has quietly dropped the first tenet of hard-core behaviourism, the hypothesis that stimulus-response learning accounts for everything. Behaviourists had bet the farm on this mechanism, but too many experiments proved that much more is involved. Furthermore, almost every researcher has drastically modified the second tenet of behaviourism, that scientists should study only observable inputs (environmental factors) and outputs (behaviour). Again, there's now too much evidence that learning, memory, representation and flexible problem-solving are all involved. As a fallback position, in effect, scientists who have conceded these two issues employ 'cognition' as a catch-all phrase. They now view animals as dealing with 'representations' that are then 'stored' and 'processed' and 'internally mediated' in one way or another. Here again we see the influence of the computer sciences grafted onto old-style behaviourism.

The following equation from the textbook *Cognitive Ecology* (not 'Cognitive Ethology') strikes me as an appropriate illustration of this general approach to animal behaviour. The subject here is the foraging habits of predators in a predator/prey relationship:

$$R = \frac{S \sum_{i=1}^{m} D_i [1-(S/M)^{k_i}] a_i^{1/k_i} e_i - (f+bS)}{1 + S \sum_{i=1}^{m} D_i [1-(S/M)^{k_i}] a_i^{1/k_i} h_i}$$

This equation may well represent terrific work; nor does working in this vein preclude holding the same views as Donald Griffin. I only include it as a striking illustration of the kind of research now directed at understanding animal *cognition*. The key point is that cognition (as the term is used in the cognitive sciences today) does not necessarily entail subjectivity or consciousness. Behaviourist-oriented researchers will grant 'cognition' to many animals, while withholding consciousness.

Now we need a definition of consciousness. *Cognitive Ethology* says this: 'An animal may be considered to experience a simple level of consciousness if it subjectively thinks about objects and events.'

Thomas Nagel rephrases this neatly: 'An organism has conscious mental states if and only if there is something that it is like to be that organism.' English philosopher John Locke, writing in the seventeenth century, had a neat little definition of consciousness:

'. . . the perception of what passes in a man's own mind'. Only if there is a subjective character to the experience, if it feels like something to be a bug or a bat or a bonobo. Nagel's famous question 'What is it like to be a bat?' implies that it's like anything at all. He is assuming that the bat does indeed possess consciousness. Then there is *self*-consciousness, or self-awareness, which adds another level to the progression. We humans have self-consciousness. Some entity I call 'I' sometimes leans back in the chair, puts its hands behind its head, and thinks about its own thinking. Do animals have such self-consciousness?

Many of the ethologists and cognitive psychologists who begrudgingly concede the possibility of some kind of mechanical 'cognition' still save 'consciousness' and 'self-consciousness' as higher levels of experience to which only we humans and perhaps the higher primates are privy. Thus the third major tenet of behaviourism, that subjective mental experiences should be ignored, is still alive and well. Some major texts in ethology do not even list 'consciousness' in the index. *The Oxford Companion to Animal Behaviour* has no such entry. Griffin and his allies think of this modified behaviourist position as 'cognitive behaviourism' – a ploy that still denies animals the essential ingredient, consciousness, by arguing that whatever the animal is experiencing, it must be a far cry from what we experience, by definition. But no one is suggesting that any animal's thoughts are as complex as a human's. As Griffin quips, the distinction may be between my three dogs enjoying a tasty treat and my studying the Battle of Gettysburg. Nor is anyone suggesting we will ever understand animal thoughts in their entirety, any more than we can understand our own. But, as the philosopher Daniel Dennett writes, we can join Donald Griffin in *avowing* the consciousness of the animal without being able to *describe* its nature in full detail.

Drawing the Lines

To we humans who enjoy both cognition and consciousness, both seem as one. Can we really grasp the idea that another animal might have just cognition without consciousness, or is this distinction just a spurious game of semantics played by the diehard behaviourists? The question is vital. If cognition is tantamount to consciousness, a lot of the issues discussed in this book would be irrelevant. There wouldn't be a debate. But the behaviourists are correct in arguing that cognition does not necessitate consciousness. I have mentioned driving a car: lots of cognition going on, but often more or less unconsciously. However I don't think driving is the perfect example, because while driving the car we can become conscious at a moment's notice. There is a better way to demonstrate that it is possible to perceive with the senses, and to act on those perceptions, and yet to be completely unconscious of doing so. The phenomenon in question is blindsight, a fascinating subject in its own right. It has been investigated in numerous tests and studies, and interpreted in many different ways.

Blindsight victims have suffered one of several forms of injury to the part of the brain known as the visual cortex, resulting in a large blindspot in the field of vision. In those cases in which both the left and right visual cortices have been damaged, the individual is completely blind. But some people with this condition can nevertheless perform quite well – much better than chance – on tests involving the identification of simple shapes and flashes of light. These patients apparently see, and then report what they see, but they have no consciousness of seeing. They have no idea how they score better than chance on the tests. They don't understand any better than the experimenters. Daniel Dennett has his doubts about blindsight and suggests that it could actually be some kind of residual vision. Maybe he's right, but I find the phenomenon useful simply as an aid for picturing cognition without consciousness.

Almost all of us 'draw the line' somewhere between those creatures we intuitively believe are endowed with consciousness, and those who aren't. A survey reported in detail in *Anthropomorphism, Anecdotes, and Animals* (edited by Robert Mitchell, Nicholas Thompson and Miles Lyn), a consistently compelling anthology, reveals that most people draw the line for basic consciousness between the invertebrates and the vertebrates, and then draw another line for some sort of higher consciousness, if not exactly self-consciousness, just beneath the dolphins, primates, cats and dogs. But we are not necessarily logical in our distinctions. Most of us relegate pigs to the lower category of consciousness, even though pigs are just as intelligent as dogs on any animal-IQ scale. One could conclude from the survey that it pays to be a common household pet rather than a denizen of the farmyard, although such a poll might yield different results since the release of the movie *Babe*. That bright piglet lifted the image of his entire species.

Thomas Nagel took into account our propensity to draw lines when he was working on his famous essay. He explained, 'I have chosen bats instead of wasps or flounders because if one travels too far down the phylogenetic tree, people gradually shed their faith that there is experience there at all.' Daniel Dennett, the other philosopher who has entered our story, wryly observes that it can be tough to draw the line even within our own lives! Was I conscious when I put the newspaper in the refrigerator and threw away the orange juice? Or when I threw the dirty socks into the toilet instead of the laundry basket? Hard to say.

Dennett proposes the following clever question to ask in order to determine where you are intuitively tempted to draw the line for a high degree of consciousness: With what creatures would you feel comfortable using the first-person plural 'we'? Would you say regarding your pet dog, 'We are going for a walk round the block'? I think you most likely would. Would you say this regarding your horse? Maybe, but to me, this seems like more of a stretch, for some reason. What about you or your child's pet gerbil or parakeet? Now we're losing a lot of folks, I'm pretty sure. Your reticulated python? Very few of us are comfortable with 'we' in this case. An oyster? That's where I draw the line!

The foraging dance of the honeybees is more sophisticated in some ways than any such organised means of communication ever observed among any other species on earth, including all the primates except *Homo sapiens*. As Donald Griffin notes, if this complex system employed by the bees were discovered among a troop of monkeys in the jungle, it would be interpreted very differently by scientists and laypeople alike. Why? Most of us do not want to attribute to honeybees the same consciousness we attribute to monkeys. We are suspicious of the role of instinct in such behaviour, and we intuitively distinguish between instinct and thinking and consciousness.

Historically, biologists have also drawn a line between instinct and some higher faculty. They have assumed that only learned as opposed to instinctive behaviour could possibly manifest consciousness. At least learning *might* require some versatility and adaptability, while it's too easy to envision instinctive behaviour, no matter how complex, as 100 per cent unconscious. As we shall see, Donald Griffin and the cognitive ethologists challenge this assumption.

On the other hand, those of us who believe that many animals do think and do have consciousness have to be careful about assuming that complexity automatically implies one or both. If this were the case, every cell on the planet would be endowed with consciousness, because every single one of them is marvellously complex. (As it turns out, panpsychism does hold that all life forms, and perhaps even all matter, possess mental states, but this is a religious or metaphysical doctrine, not science.) So, finely tuned complexity proves nothing. Learning may not be proof positive, either. Primitive invertebrates with very simple neural networks are capable of relatively astonishing results. One particularly well-known example within biology is the earthworm, which was studied carefully for decades by none other than Charles Darwin. The earthworm draws a leaf into its tunnel by grasping its narrow tip, rather than the stem end. This is the most mechanically efficient way, without a doubt, but is this the kind of action that instinct could instil? It's not easy to accept this, and Darwin didn't. But if the action is not instinct, does the earthworm *learn* the most efficient way to proceed? This seems to be the only alternative. But do we want to endow the earthworm with consciousness? Most of us draw the line a little higher – but we could be quite wrong.

The Law of Parsimony

In all scientific investigation, the law of parsimony gives priority to the simplest possible explanation. Start with the simplest explanation, and if it 'works', stop right there. If it doesn't work, consider something a little more complicated. In biology, the corollary known as Morgan's Canon states, 'In no case may we interpret an action as the outcome of the exercise of a higher psychical faculty, if it can be interpreted as the outcome of the exercise of one which stands lower in the psychological scale.'

If a given behaviour can be plausibly explained as instinct without need of cognition, draw the line; if information processing – cognition – seems sufficient, then there's no need to invoke conscious thought, so draw the line.

C. Lloyd Morgan was a British psychologist, a colleague of G.J. Romanes, Darwin's protégé. With his Canon, Morgan was essentially reacting against Darwin's and Romanes' excessive anthropomorphism. Promulgated to a less-than-waiting-world in 1894, the decree is now viewed as marking the end of the era of systematic anthropomorphism. His Canon is sound enough policy, but it was often either misinterpreted or misrepresented as a defence of strict behaviourist explanations. In fact, Morgan accepted some degree of animal consciousness, as he wrote in *The Limits of Animal Intelligence*. He even amended the often misunderstood Canon with the following statement: '. . . it may be added – lest the range of the principle be misunderstood – that the canon by no means excludes . . . higher mental processes if we already have independent evidence of their occurrence.' Morgan did not doubt that a dog has a juicy bone literally 'in mind' when he looks wistfully at his master working over the stove. Or, as Stanley Coren writes, in *The Intelligence of Dogs*, the Canon, if properly invoked, distinguishes between the intelligence required for a dog to find his way home after taking inventory of all the fire hydrants in the extended neighbourhood, and the intelligence required to read a map of the streets. Dogs have the one talent but not the other.

In the Clever Hans episode, the higher-level explanation – counting – substituted for a more plausible 'lower' explanation. This was a violation of Morgan's Canon, and it was found out. But then Clever Hans' failure to do arithmetic was taken as evidence that he was incapable of any thought whatsoever. But couldn't Hans have been thinking about the inadvertent cues of his owner and the other testers? Couldn't he have learned, as Donald Griffin puts it, 'I must tap my foot when that man nods his head'? This wouldn't be arithmetic, but it would be some kind of thought.

In the first chapter we met primate researcher Frans de Waal, who convincingly suggested that it is impossible to look into the eyes of a higher primate and not find 'the age-old connection'. One of the chimps de Waal had in mind was Georgia, a resident at the Yerkes Field Station in the state of Georgia. This mischievous animal has the habit of surprising human guests to her quarters with a sudden spray of well-aimed water. She does this all the time. If necessary, she'll wait with tight-lipped tension until the victim comes close enough for the prank to have its drenching effect. Other observers have reported similar behaviour with other chimps in other facilities, and it could be interpreted as affection or aggression or anything in between; for my current purposes, the specific interpretation doesn't matter. In his foreword to *Anthropomorphism, Anecdotes, and Animals*, de Waal describes how he once succeeded in 'facing down' Georgia, who he knew was sneaking up on him with a mouthful of water. He writes:

> I looked her straight into the eyes and pointed my finger at her, warning, in Dutch, 'I have seen you.' She immediately

stepped away and let part of the water drop, swallowing the rest. I certainly do not wish to claim that she understands Dutch, but she must have sensed that I knew what she was up to, and that I was not going to be an easy target.

Any human who acted as Georgia acted with regularity would have been scolded and held accountable, while any animal, even the one closest to us in genetic endowment, the chimpanzee, is still considered by many investigators, in de Waal's phrase, 'a mere passive instrument of stimulus-response contingencies'. This is absurd, in my opinion – a classic example of what the primatologist Roger Fouts calls the 'rubber ruler', with which the explanation depends on which creature is manifesting the behaviour. Latter-day behaviourists who withhold consciousness even from chimps – and some do – forget that the second part of Morgan's Canon 'by no means excludes . . . higher mental processes if we already have independent evidence of their occurrence'. In the case of chimps, we have this evidence in abundance.

Onward, Through the Fog

A professional trainer of big cats once adopted a Rhodesian ridgeback pup and a lion cub, who then grew up together. The dog, a female, was named Janee, and the lion, a male, Wazoo. They were inseparable, night and day, even sleeping together while rolled up in a ball. Somehow, Janee succeeded in convincing the lion that she was the leader of this 'pride' of two. She would snarl, bark and occasionally even nip the lion on the ear in order to bring him to heel. Both animals gave evidence of separation anxiety if they were apart for any period of time. But then, after seven years of this incredibly close relationship, and within the space of a single week, matters changed. The lion suddenly woke up to his role in the larger scheme of things. He realised he was a lion and Janee was a dog. One day, as Janee tried her usual dominant behaviour, Wazoo snapped and charged her. Their owner saved the day, but that was the end of a beautiful friendship.

This story is from the psychoanalyst Jeffrey Masson's *Dogs Never Lie About Love*. Masson asks in conclusion, 'I wonder if either animal is ever nostalgic. Does the dog think back with longing for the day when her closest friend was a lion? Does the lion ever wonder why he spoiled a beautiful relationship?'

It's great to think so, I believe. The story is also a neat exercise in drawing the line. Some might feel the dog remembers the lion, but not vice versa. But that only underlines the human bias towards dogs. Anyway, even scientists can now entertain such questions without a guilty conscience, because the cognitive revolution has opened the door wide to theories and experiments and field work regarding animal emotions and thought and consciousness. What specific kinds of behaviour would plausibly infer mental states and therefore give us something further to investigate? This question has been worried and debated to death, as we should expect. Donald Griffin proposes two over-arching categories: adaptive behaviour and interspecies communication. (His third source of inferential evidence is the physiology of the brain.) Here's a list of other candidate behaviours culled from the literature:

- Complexity
- Learning from the past in a way that can be applied to the future
- Plastic memory
- Focused attention
- Concept formation
- Category formation
- Tool-using
- Tool-making
- 'If-then', or 'purposive', thinking and behaviour
- Deceptive behaviour
- Use of symbolism

Beginning in the next chapter, we'll look in detail at a wide range of behaviour from the laboratory and the field that suggests intelligence. I will be careful to note challenges to these assumptions as well, because every issue in cognitive ethology is open to challenge. In fact, this is not a particularly good subject for those who want sure, easy answers. On the other hand, there's no better subject for fascinating questions, because the world of animals is infinitely rich and infinitely puzzling. Just like our own.

chapter 4

MENTAL MAPPING

From Babies to Birds

In the middle decades of this century, the famous Swiss psychologist and historian of science Jean Piaget developed his enormously influential model for cognitive development in children. Piaget argued that children learn in a genetically programmed way, following clearly defined stages which usually arrive, on cue, at specific ages. Piaget hypothesised, and got very close to proving, that these stages are, in effect, mental reflexes. According to his theory, learning is just as instinctive as breathing, and we do both, automatically.

This depiction of 'mental mechanisms' interacting with the environment was very influential among ethologists and comparative psychologists exploring animal behaviour. Indeed, Piaget himself was also interested in these implications; in 1970, he included his own extension of his theories regarding animal cognition in his magnum opus *Biology and Knowledge*.

An early and essential stage of learning in children is what Piaget labelled 'object permanence'. This means learning that objects can stay put, and then remembering *where* they stay put. The infant has to put together a map – and then many maps, every day. Children acquire their first rudimentary ability in this area between eight and 12 months and finally achieve complete mastery – 'stage 6' – between 18 and 24 months.

It turns out that quite a few birds and mammals, including some of the 'lower' primates (like squirrel monkeys), achieve one of the intermediate stages of learning object permanence. And some animals achieve total stage 6 competence. Cognitive ethologists often refer to this general category of learning in animals as 'mental maps'. (As we saw in Chapter 2 , behaviourist researchers refuse to use the term 'mental' – they study 'cognitive maps' instead. But mental maps and cognitive maps are just different labels for the same ability. The issue is one of interpretation.)

The simplest tests, studying the simplest kind of spatial memory, date back to at least the 1930s. In these experiments, two cups are placed in front of the animal – a chimpanzee, say. One of the cups is baited with food in clear view of the chimp. Then a screen is lowered. After a delay, which varies from test to test, the screen is raised and the subject is free to 'search' for the food. About 90 per cent of chimpanzees accomplish this task easily. They proceed straight to the baited cup. Macaques score lower; only about 60 per cent go immediately to the food. Since no sensory cues are guiding the animals at the time of the search, researchers have concluded that the successful chimps have created a basic 'map'. Moreover, chimps will still proceed directly to the baited cup even if the set-up is rotated slightly to the left or the right.

Such experiments were among the first to challenge the behaviourist bias, according to which learning would take place only as a corollary to immediate stimulus-response conditioning and gratification of immediate needs. Strict behaviourism had a hard time explaining the ability to remember a mental map; it also had a hard time dealing with the animal who can 'produce his response in the absence of the stimulus', as Jacques Vauclair writes in his invaluable book *Animal Cognition: An Introduction to Modern Comparative Psychology*.

Sue Healy and her colleagues from the University of Newcastle have spent a long time in the Rocky Mountains, trying to determine the mental mapping abilities of the rufous hummingbird. Why does Healy study this particular bird? Mainly because she can. These beautiful little birds migrate north from Mexico, all the way to the Canadian Rockies, for about six weeks of mating activity in the late spring and early summer, before returning home. Male rufous hummingbirds, easily identified by their ruby-red throats, are very aggressive defenders of their territory. They feed in a flower for only a few seconds before returning to a perch to check for potential rivals in the mating game. It is therefore fairly easy for Healy to attract just one bird to her experimental set-up and then be certain that it is this same bird who has returned a second time, as required for the test. This kind of experimental control is difficult to achieve with most birds, for obvious reasons.

Does the rufous hummingbird in its springtime mountain habitat have the same (or even a superior) mapping ability as the chimp in the lab working with the cups hidden behind the screen? Does the bird remember which flowers it has already visited and depleted of nectar? Healy thought it might, because such feeding efficiency would be a tremendous advantage, saving both time and energy on the arduous migratory trip. She explains, 'A typical territory has about 200 flowers. When the bird empties a flower of nectar, the estimated time before the flower refills is four hours. Rather than coming back to a flower and finding it empty, it would be useful for the bird to be able to remember it and avoid it.'

Healy's experiment is fairly straightforward. First, she sets up a semicircle of eight artificial flowers, which she calls a radial maze. The male rufous who finds this maze to his liking becomes the unofficial master of this domain. Healy watches to make sure, of course, but she knows that this one bird will be the subject of her experiment, because he'll drive away any potential transgressor. Anyone who has watched the behaviour of hummingbirds in these mountains can testify to this. The male rufous is a domineering sort, no doubt about it.

The flowers are baited with very little sucrose, so that a given flower will be depleted *before* the hummingbird abruptly leaves it for another. Since the idea is to determine whether the bird remembers that the flower is empty (and therefore doesn't waste time and energy finding out on a second visit), this depletion is a vital stage of the experiment. The subject bird is allowed to feed at only four of the eight flowers. Then he is shooed away and Healy sits quietly, waiting for him to return, which he almost always does, usually within ten minutes. On the return visit, more than 75 per cent of the hummingbirds have flown directly to the four flowers they had *not* visited the first time. They did indeed remember the flowers previously visited and depleted, and avoided them.

Maybe a 75 per cent 'pass rate' doesn't sound that good. But one of Healy's colleagues back in Newcastle demonstrated that about the same percentage of college students are successful on an equivalent test using overturned cups and sweets. And the bird's brain is 7000 times smaller than the brain of an undergraduate! Healy always laughs when she tells this story.

But how do the hummingbirds remember the depleted flowers? Do they remember a particular route flown among the eight flowers, or do they remember the specific locations of the four? Further tests seem to prove that it is locations, rather than routes. Or could it be some identifying characteristic of the particular flower? There is a huge difference. Remembering location indicates the presence of a mental map; remembering colour and shape is also a significant feat, but it is not a mental map.

To address this possibility – that the birds are remembering specific flowers, not specific locations – Healy moves onto the next stage of her investigation, in which she sets up an array of distinctively and individually designed flowers, only one of which has sucrose. The others have just water, which hummingbirds don't like at all. (The birds get all the water they need from nectar. Drinking plain water is a waste of time, and they don't do it.) When the rufous has found the one flower in Healy's radial maze which has been filled with sucrose, she shoos him away *before* the flower is empty. This is key. Unlike the previous test, this one is set up to entice the bird back to the same flower.

After the bird has been temporarily banished, Healy switches the array before his return visit. About 80 per cent of the time, the returning bird flies directly to the flower in the same location as the originally baited one (not to that flower itself, which is now in a new position). If Healy does not place any flower in the target location, the bird flies there anyway and circles around, searching for the flower where he knows it should be. This test confirms that the bird remembers location, not specific flowers.

Scientists have confirmed that hummingbirds discriminate very well between different colours. Why don't they use this talent for feeding purposes? The answer seems to be simple. Most of the flowers in any given territory will be the same kind, and almost identical in colour, so colour would be a dicey basis for discrimination. And, as Healy puts it, 'flowers in real life don't move'. (Interestingly, in the 20 per cent of cases in which the bird did not return to the correct location, he selected the original target flower in its new location. This seems to indicate that colour remains a secondary consideration that occasionally takes precedence, for whatever reason.)

The Clark's nutcracker may be the world's most studied bird in terms of mental mapping ability. This bird, which inhabits the mountain pine forests of the American Southwest, undoubtedly employs a detailed map of its terrain while harvesting pine seeds in a period of just a few weeks in the late summer and early autumn. It then stores them in up to 2000 locations. The usual nutcracker cache holds two to five seeds but a single bird may store up to 33,000 seeds when the cone crop is heavy. Students of this nutcracker estimate that a bird must be able to retrieve 3000 seeds in order to survive the difficult alpine winter.

In laboratory tests by Vanderwall, Balda and Turek, the nutcrackers were presented with 18 cache sites, which they proceeded to use for storing seeds. Then the birds were removed from this experimental environment for 11 days. In further laboratory research with the same birds almost a year later, they were as accurate at finding their caches after 285 days as they were after 11 days. A spectacular feat of memory! As Dr Russell P. Balda, of Northern Arizona University, commented, 'Not even humans can perform at the accuracy levels that we see in these birds.' (This comes as no surprise to me because I can never remember where I've put my keys.) The result confirmed that the birds were using a mental map, as did the results when landmarks within the lab were shifted around after the initial caching operation. Then the nutcrackers' scores plummeted.

The landmark thesis was beautifully proved in another set of experiments with chickadees. These birds were trained to feed at one of four boxes attached to a wall of their aviary. Then the array of boxes was shifted laterally on the wall. The shape of the array stayed the same, but now a different box was in the position of the original feeding box, relative to the landmarks in the room. When the birds were returned to the aviary with this new set-up on the wall, their first response was to try the box in the same position on the wall as the original feeding box; second choice was the box in the same position relative to the array as the original box; third choice was the box with the same identifying marks as the original box. The only possible conclusion was that landmarks in the overall environment – maps, in a word – are the most important information used by these birds.

Researchers have long understood that birds have a talent for simple mapping, but it was only quite recently that they learned – to their great surprise – that certain birds remember not only where they stored food but *when* they stored it. The experiments leading to this conclusion involved scrub jays which were given both waxworm larvae and peanuts to store. These larvae, usually the jays' preferred food, rapidly lose their desirability over time. When the jays were then given the choice of which food to retrieve, they chose the larvae only if they had stored the worms quite recently. If they had stored them several days earlier, they chose the peanuts instead. Clearly, the scrub jay remembers when it stores food and then takes this memory into account when choosing its menu. This is a remarkable experimental result. Until this research was first published in late 1998, this kind of 'episodic memory' was considered beyond the capability of most, if not all, animals. It was a talent that separated us humans from the hoi polloi, as *The New York Times* phrased it. But if the scrub jay remembers when it buries food, why shouldn't other birds and animals?

Of course, no avian feat of memory can match the navigation skills of migratory birds. They are unmatched by any creature. The basic map-and-compass system is not the half of it. Star-reading is necessary for these fliers, and magnetic fields may play an important role. Long-haul migratory birds, such as the arctic tern (which commutes annually between the two polar icecaps, a round-distance of more than 25,000 miles), easily solve navigational problems that bedevilled human navigators right up until this era of the Global Positioning System. It is no exaggeration to say that migratory birds do not need GPS. They have their own – but exactly how it works, no one knows.

The Case of the Kidnapped Bee

Many insects are successful in the earliest stages of learning 'object permanence' and creating mental maps. Once we get past the fact that they are insects (creatures we tend to think of as essentially instinctive), we really shouldn't be surprised by this mapping ability, especially in those species that feed on flowers. If mapping is advantageous for foraging birds, why shouldn't the same be true of flower-foraging insects? Perhaps mapping is just too advantageous *not* to have evolved among species, be they birds or bees.

Bumblebees are a case in point. G. H. Pyke has studied the food-gathering behaviour of *Bombus appositus* while foraging for monkshood flowers in Colorado. Pyke tags his bees, then films them as they invade a succulent new 'garden'. In general, the bees start with the flower at the bottom of a given plant and work their way up the shaft until they reach the top, when they shift to a neighbouring plant. Pyke learned that these bees, like the rufous hummingbird, almost never waste time and energy returning to the same flower twice. They remember, for a short time, at least, which flowers they have visited, or they leave a scent that serves as a reminder, or both. In any event, as Donald Griffin puts it, in *Animal Minds*, these are not the 'simple, stereotyped sort of reactions we are accustomed to expect from insects'.

Extremely sophisticated computer analyses of the minutely observed movements of ants support the case for some kind of mapping ability in these creatures. (As ground dwellers, ants are more amenable to round-the-clock observation than wasps and bees. Reports of this research are full of equations, and phrases like 'trigonometric vector summation'.) The latest thinking holds that all ants and bees employ some combination of a sun-based compass and a magnetic compass, as well as the more mundane expedient of landmarks and mental maps.

This landmark hypothesis was first conclusively proved by Nikolaas Tinbergen's classic 1950s experiment on the homing instincts of the digger wasp, *Philanthus triangulum*. Tinbergen's idea couldn't have been simpler. First he arranged a set of pine cones around a wasp nest that was a centre of activity. When most of the wasps were away from the hive, he then moved the cones into the same relative position around a fake nest. The wasps all returned to the fake nest.

This was an incontrovertible experiment, replicated many times and with many modifications. But it was not the last word on the subject, by any means. Consider the study first reported in 1986 by J.L. Gould, Professor of Ecology and Evolutionary Biology at Princeton University. I read about his work, which might be called 'The Case of the Kidnapped Bee', in quite a few different sources, including *The Animal Mind* (published in 1994), which he co-authored with Carol Grant Gould, a science writer.

The study involved a number of different experiments. In one of these, the bees were first trained to follow a route from their hive to a food source due west – about 140 metres (150 yards) away. Then one bee was kidnapped, while *en route* to this food, and carried in an opaque box to a new site due south of the hive. The two sites – one to the west of the hive, one to the south – were not visible from each other, but in some of the trials there was a landmark tree in the vicinity. In the trials with the landmark, most of the kidnapped bees flew directly from the new release point (south of the hive) to the old site (west of the hive), without proceeding to the hive itself for orientation purposes. This route was one the bee had never flown before. Kidnapped bees which were released only 30 metres (33 yards) further from the landmark tree than the first bees found it harder to pick out the correct route to the old site. Some became disorientated, while others who did fly off towards the goal did not follow quite as straight a line as bees released closer to the landmark tree.

A lot of experimental energy has been spent trying to replicate (and explain) these 'mental mapping' results, which are rather more complicated than my summary might imply. However, the consensus is that landmarks are definitely important, if not the whole story, and that the bees use the landmarks to construct some kind of mental map.

As Gould writes:

> The complications inherent in testing the cognitive abilities of unrestrained animals in the wild, particularly when they have more than one way to solve a problem, is well illustrated by the bee-map saga . . . Despite a widespread unwillingness to take indications of insect intellect at face value, there is good evidence that even a few milligrams of highly specialized neural wiring can accomplish a limited set of individually impressive cognitive tasks essential to the natural history of the animal in question.

Gould notes that, in comparison with flies and moths, social insects (like ants, wasps and bees for example) have excess brain power relative to body weight, and he adds, 'The excess brain power of social insects may well have evolved to solve problems based on goal-oriented strategies — strategies that require a level of cognitive finesse far beyond anything imagined in invertebrates by most researchers as recently as two decades ago.'

The Maze

The maze has been standard equipment in the laboratories of comparative psychologists since the nineteenth century. In fact, the very first experiments that challenged behaviourist principles were those in which lab rats were turned loose in a maze while they were neither hungry nor thirsty. According to theory, they therefore had no reason to learn anything. But they did. In fact, Piaget was right: learning is automatic. It is what we animals do, at whatever level we're capable of doing it. The maze is also perfect terrain for testing mental maps under rigidly controlled conditions. Therefore, it's not surprising that a great deal of work has been done in this area with laboratory animals.

In 1993, Chapuis and Scardili showed that hamsters could build a mental map of a maze, even if they had only been in half of it! And for the *Nature* series on the animal mind the team recreated this classic experiment with the favoured laboratory rat. The set-up was similar to Sue Healy's radial maze of flowers, but in this case the maze was shaped like a cartwheel – with rim and spokes. In the learning part of the test the rat was allowed into only half of the perimeter; the rest of the maze was out of bounds and blocked off. Food was placed at the end of one of the arms and the rat quickly learned where to find it. It ran straight to the food, *by the shortest route*, every time it was placed in the maze. How did the rat know which route to take?

The obvious answer would be smell. The rat may have simply smelled its way to the food. To test this possibility, food was placed at the ends of *all* the arms. If the rat was working on smell alone he should have gone to the nearest food source. But the rat wasn't using smell. It was using its memory. As far as the rat knew, there was food only at the end of one arm. Using its mental map of the maze, it worked out the best route to reach it. Thus, rats can work out alternative routes, just as a London cabby would if he found a street unexpectedly blocked. A London cabby's astonishing memory is an extraordinary example of mental mapping. But perhaps the rats in the sewers beneath London's streets are doing the same thing.

As we should expect by now, complicated, behaviourist-oriented explanations have been propounded to explain all these results. These have included 'habit family hierarchies' and 'fractional antedating goal responses', which last incomprehensible theory prompted one commentator to write, '[this] system seems to require more cognitive effort by both the human theorist and the rats'. Meanwhile, the law of parsimony (introduced in the previous chapter) proposes that the simplest, most straightforward explanation stands a good chance of being the best one. E.C. Tolman, an important researcher in this field, has written, 'We believe that in the course of learning, something like a field map of the environment gets established in the rat's brain and the incoming impulses are usually worked over and elaborated into a tentative, cognitive-like map of the environment.' (Tolman knew early in this century that any number of maze tests had already disproved the strict behaviourist model; but he was intimidated by the behaviourist influence, as was almost everyone else. Therefore he coined the term 'purposive behaviourism' – a classic case of hedging one's bets. Later in his career, he was more candid.)

Psychologist John Pearce, of Cardiff University, agrees with Tolman. Pearce thinks of a mental map as something like an aerial photograph, based on landmarks and relationships. And he says he 'can be reasonably confident' that his laboratory rat is finding the food in the maze 'by use of a cognitive or mental map, rather than by a sense of smell'.

It turns out that turtles (not the brightest of species by most measures) can also learn a complex maze after just a few trials, proceeding not just slowly but methodically as well, 'visually scrutinising each choice point', as the Oxford Companion to Animal Behaviour puts it. Chameleons also happen to be very good at navigating complex mazes and solving related situations known as detour problems. Astonishing? Not at all. In their natural habitat, chameleons are, according to the Companion, 'frequently required to make detours among the branches of trees to come within striking distance of their prey'.

And chimpanzees are also excellent with a mental map, although I doubt whether they could replicate the performance of the Clark's nutcracker. (Why not? Because they don't need to!) Young chimps and gorillas master object permanence even more quickly than human children do; but this is not as puzzling as it might at first seem, since young primate babies mature more quickly in most respects than humans. They can move around in their environment after just one month, compared with eight or nine months for a human baby, so they can start learning about the wide world at an earlier age.

One of the best-known chimp experiments was first set up by E.W. Menzel in the 1970s and has now been repeated by many other primatologists. The subject chimp is carried around a large outdoor space by one experimenter, allowing the animal to observe a second experimenter hiding food in randomly selected sites. This tour of the area takes about ten minutes, after which the chimp is returned to the company of his or her mates, and all of them are released into the test zone. The movements of the test chimp are closely observed.

In a way this experiment resembles the famous travelling salesman problem that continues to fascinate mathematicians, in which the aim is to find the shortest route that will take the salesman to every potential customer on his map. If there are a lot of customers, super computers are challenged by this task. The chimpanzees do not solve their particular problem perfectly, but they perform very well, retrieving an average of 13 out of 18 items of food. More importantly, they do so efficiently – not simply following the route taken initially, but using short cuts and moving directly from food cache to food cache. Menzel concluded, 'Especially since a novel set of locations was used on every trial, and the informed animal ran directly to hiding places that were not visible from the point from which their run commenced, the term "cognitive mapping" seems as good a descriptor as any.'

Other experiments, both in controlled conditions and in the forest, have confirmed Menzel's first results. One of the most intriguing of these tests utilises closed-circuit television. In the experiment, chimps watch on television as an experimenter hides somewhere in an enclosure. In effect, the television screen is a map. Can the chimp read this map? Yes. Released into the enclosure, most chimps proceed directly to the hiding human. They can also relate the contents of a 'doll's house' to the equivalent full-size room.

Are the chimps, the chameleons, the lab rats, the bees and the birds all mapping their terrain in the same way? Are they mapping their terrain the same way we do? It's difficult to say. Bees don't seem able to operate effectively over as great a distance as birds. At several hundred metres, bees get lost if they have to find a novel route. The nutcrackers, in contrast, can cache and retrieve their pine nuts over immense areas of forest.

Rotation

A corollary of the 'mental maps' capability is the ability to recognise and manipulate mental images. For example, honeybees can recognise a mirror image of a flower-like shape. (This has been proved by removing a target flower and substituting its mirror image and a choice of other patterns. The bees will always prefer the mirror image.) The very idea that bees could remember images in this way was heretical as late as the 1970s. But it is a fact. Even more impressively, bees can perform a 'mental rotation' of a two-dimensional object.

Pigeons turn out to be excellent in a variety of tests that require the manipulation of mental images. In one classic experiment, the hands of a clock displayed on a computer screen disappear and then reappear while moving at a non-constant rate of speed. Pigeons are nevertheless able to accurately estimate when and where the disappearing hand will reappear. This is not easy to do, and the birds have been subjected to a battery of finely tuned tests which try to pin down how they're so successful. The best explanation to date is that they consult some kind of mental imagery. Pigeons are also able to recognise images rotated any number of degrees the experimenter chooses. (This work was done by Nieworth and Rilling in 1987.)

What's really fascinating is that pigeons apparently do not do what human subjects do in these tests, which is to rotate the image in the 'mind's eye'. (We know that humans do this 'mental rotation' because the time required for us to solve the problem is directly correlated with the degree of rotation.) Jacques Vauclair, a specialist in this area, showed that baboons also excel in these rotation problems and apparently solve them as we do, by 'mental rotation', because their times also vary with the degree of rotation. (Of course, their times are faster than ours by a factor of up to eight! Vauclair is not sure why this is, but he hypothesises that we get bogged down asking extraneous questions, perhaps unconsciously.) Studies with dolphins indicate that they, too, adopt a humanlike mental strategy. The pigeons' response rate, on the other hand, is constant, no matter how much the image has been rotated. Experimenters therefore conclude that the bird is doing something different from us; something pre-wired, presumably. Oh, and one more point: they're also about three times faster than we are on these tests.

Slow But Sure

We think of elephants as denizens of the bountiful savannahs of East Africa, but a small number – about 200 – live in the Namibian desert of southwest Africa, one of the harshest environments on earth. It is startling to see these immense animals trudging slowly over baked desert floors, alongside barren sand dunes. We feel as if we're suddenly on a different planet – the wrong planet. But here they are. It is also startling to realise how big these elephants are – bigger, on average, than their counterparts elsewhere on the continent, although we might imagine that the environment would dictate just the opposite. In fact, the largest African elephant ever measured – 4.2 metres (13 feet, 10 inches) – was found in this desert.

Where is the water around here? Often enough, the water is a long way from the best feeding grounds, as much as 64 kilometres (40 miles) away and several feet underground as well. If these elephants merely wandered around the desert aimlessly seeking both food and water, they would find neither and they would die. There would be no desert elephants of Namibia. But the elephants know where all the water is, where all the food is, and where all the best routes between water and food are. This knowledge is not instinct. They have learned the map of the region and passed it down through the generations. If we do not eradicate them with our rifles and our encroachments into their homeland, they will survive.

chapter 5

CATEGORIES AND COUNTING

Raising the Stakes

In his book *The Others: How Animals Made Us Human*, Paul Shepard writes about the prodigious naming and categorising habits of the few remaining tribal societies around the world. He quotes one anthropologist who estimates that such societies incorporate in their everyday working vocabularies the names of over 1000 plants and animals. Nor do these people need to know all these species for economic reasons (any more than most stamp or coin or beetle collectors need their vast knowledge for economic reasons). We humans collect and categorise things for the sheer pleasure and ancient purpose of it.

Can animals establish categories or, as they are labelled by researchers, 'learning sets'? We have seen how all kinds of animals use mental maps in order to get around the world and forage for food. This cognitive ability is so widespread, it appears to be almost an entry-level requirement in the wide world of animal intelligence. Cognitive ethologists agree, and raise the stakes; they look for learning which requires a degree of abstraction – a tougher proposition than working only with objects such as landmarks and food hidden under cups. If an animal can learn one category, can it then transfer this learning to the conditions in another category? In short, can an animal learn how to learn?

The answer is yes.

Let's begin a brief survey of what has become a mountain of evidence by looking at a set of experiments with macaque monkeys, generally considered by primatologists to be 'middle range' primates in terms of intelligence. First, the monkey is trained by rewards to discriminate between two objects, say a ball and a box, and is then rewarded for selecting the ball. After this pair of objects is mastered – when the monkey has reached the point at which he always selects the ball because this selection produces a reward – a different pair of objects is introduced and the monkey is trained, again by reward, to select one of the two. After he has learned to always select the same object, another pair of objects is introduced, then another, and so on. What happens over this span of trials is quite instructive: the monkey gets to the point where, after dealing with several hundred different pairs, he is perfect. With any subsequent new pair, he picks the object for which he is first rewarded every time. Simple enough, perhaps, from our perspective, but the experiment proves that the monkey has learned what is expected of him and is able to generalise this learning in novel situations – with each new pair of objects.

A more complex test of this type is called the matching-to-sample (or MTS) test. In this experiment, one object is presented along with two or more test objects, one of which is identical to the sample. Colour is a common criterion. A red ball, say, is presented in the company of one red ball and one green ball. The idea is to pick the red ball. Food is used to reinforce the correct choice. If the experimenter so chooses, the 'opposite' category can also be trained and tested for when the idea is to pick the different object, in this case the green ball. Primates are excellent at such same-or-different tests. Pigeons can also learn this skill, but they are slower to pick it up.

The most difficult challenge of this type is the 'relations between relations' test, in which the animal is 'asked' whether a pair of red balls has the same relationship as a pair of balls, one red and one green. Such reasoning by analogy is reasoning of a high order. Children don't acquire this skill until past infancy, and when the relationship is subtle this test can even challenge adult humans. Young chimpanzees have no success at all with analogical reasoning, but adults chimps do, as amply demonstrated by Sarah, the main subject of the book, *The Mind of an Ape*, by David Premack and Ann James Premack. When asked if a small and a large circle stand in the same relationship as a small and a large square, Sarah uses a symbolic chip (whose meaning she has already mastered) to answer yes. Do the half-moon and the same shape with a black dot stand in the same relationship as the triangular shape and the same shape with a black dot in it? Yes. Do the half-moon and the same shape with a black dot stand in the same relationship as the 'home plate' and the same shape coloured completely black? No.

Amazingly to me – I am not sure how well I would do – Sarah is correct on about three-quarters of such tests. She does even better on what are called 'conceptual analogy problems', some of which would challenge us humans. For example, after being shown a lock and key (Sarah knows the relationship between the two), she is shown a closed paint can and asked to select either a can opener or a paintbrush, both of which she is familiar with. This relationship is tricky, because either implement could logically be used with the can of paint, but only one stands in the same relationship with the can as the key does with the lock. Sarah correctly picks the can opener, not the paint brush. In one trial, she was correct in 15 out of 18 similar tests. It is difficult if not impossible to see such results as indicative of anything other than intelligent thought.

Sarah also excels on the intriguing 'conservation test', which stumps most children under the age of seven. In this challenge, two identical glasses are filled with identical amounts of water. Is the amount of water in each glass the same? Yes, five-year-old children say, and so does Sarah using her symbolic coloured chips which mean 'same' or 'different'. Next, the water in one of the glasses is poured into a narrower glass. Is the amount of water in this narrower glass the same as the amount in the other glass? Most five- and even six-year-old children answer no, because, they explain, the water level is higher. These children have not mastered the concept of 'conservation' regarding water. Yes, says Sarah, well over half the time. She understands and uses the concept of conservation. Or does she?

The attribution of 'conservation' to the chimpanzee has been challenged on a variety of grounds, all coming down to the charge that she is not doing what she appears to be doing. She simply must be doing something else. Why? Because she's a chimpanzee, not a human; a concept of categories requires language, and chimps don't have language. In short, we may have here a classic illustration of Roger Fouts's 'rubber ruler': the interpretation of a given behaviour depends on the animal involved.

Those Amazing Birds

Professor John Pearce of the University of Cardiff, as noted earlier, works with rats, mazes and mental maps. Very conveniently, he is also an expert in category tests, and his favourite animal in this regard is the trusty pigeon. His long-standing work with these birds and their remarkable ability to handle categories builds on the groundbreaking work of Richard Herrnstein of Harvard University in the 1970s. It is utterly fascinating – and highly controversial – research.

In Herrnstein's categorisation test, pigeons were first presented with dozens of pictures, one at a time. Some of these pictures had a tree somewhere in the scene, while the others did not. The object for the pigeon was to learn to select the picture with a tree. When it did so, it was rewarded with a bit of food. If it pecked at a picture that did not contain a tree, the bird was not rewarded. The pigeons learned very quickly to recognise trees – even if only a leaf or branch or an aerial shot of a forest was showing in the picture. The pigeons seemed to have learned to recognise not one particular tree, but a generalised concept of 'tree'. To test this, the pigeons were shown a new batch of pictures with trees of all shapes and sizes hidden among other pictures of plants containing no trees. The pigeons caught on immediately, recognising the trees almost every time.

Perhaps it is not surprising that pigeons can recognise trees. They would be in trouble if they could not! Are pigeons hard-wired (or genetically programmed) to know trees? To test this conceptual ability, John Pearce has been conducting experiments based on some work from Japan by Watanabe. For example, can pigeons learn artificial concepts such as distinguishing between impressionist and abstract painting styles? Surprisingly, the researchers discovered that pigeons could, indeed, be taught to discriminate between paintings by Picasso and paintings by Monet. Moreover, when shown new paintings they had never seen before, the pigeons were able to classify them as impressionist or abstract.

Amazingly, Pearce has devised categorisation tests in which pigeon performance is much better than human performance. In fact, category tests that are almost impossible for Pearce's undergraduate students are easily solved by the birds. For example, in one test a series of abstract patterns was flashed on a television screen. For the students, some of the patterns were followed by a tone. For the birds, the tone was replaced by a reward of food.

The students' performance on this test was miserable.

'Oh dear, that's not very good,' their professor said. 'I don't want to worry you, but my pigeons find this problem quite easy. For those of you who have failed to match the intellectual skills of my pigeons, the solution is the average height of the pattern shown on the video screen. When the average height is low, then a tone is presented. But when it is high – no tone.'

Most of the students hadn't worked this out, but they did beat the birds when the challenge was to select the pictures in which the shapes were of equal height. The birds did poorly on this version of the test.

So what's going on here? Are the pigeons 'thinking' in the same way as we think on these tests? It's not clear, but the results do seem to indicate that pigeons are capable of learning 'relatively abstract classificatory systems', to quote Jacques Vauclair. Many experimental psychologists have their doubts, however. After all, if they challenge Sarah the chimpanzee, they're likely to challenge a flock of laboratory pigeons.

The first and most obvious possibility is that the birds are simply memorising the pictures they have pecked and for which they have received a reward of food. While we amateurs in this field might think that this would require a highly unlikely display of excellent memory by a mere bird, pigeons can in fact memorise hundreds of pictures and remember them months later. This has been proved in many control tests performed in many different labs. Pigeons have excellent visual memory. In this case, however, there is a way to find out whether the birds are just memorising the pictures paired with food. Were this the case, they should score only at chance levels when presented with a new set of pictures. With a new set, they have not had a chance to remember the rewarded pictures. If, however, they really have internalised a category of 'trees', then their initial score with this new set of pictures should be much better than chance.

May we have the envelope please? The pigeon's score with new pictures is *much* better than chance − about 75 per cent! They know what they are pecking at, and even some of their mistakes are perfectly understandable when, for example, the picture included a shape, such as a telephone pole, that might well have been a tree. Such control tests prove that the pigeons are not just remembering rewarded pictures, but have formed the concept of a tree.

But wait a minute. Maybe they have formed the concept not of 'tree' but of 'tree-like shape'. There is a big difference between the two. One is a feat of genuine cognitive or mental abstraction, the other a feat of visual acuity. Critics of the results point to the errors, such as the telephone pole, as indicating that the birds are working only with visual categories. It is correct that a telephone pole does look somewhat tree-like, so this error might indicate that the birds are responding only to visual clues − 'a small set of relevant features', as one observer phrases it − and not to any category concept we humans bring with us to the experiment.

Moreover, pigeons can discriminate at a much better than chance rate between happy and sad human faces. Do we really believe they are thereby demonstrating shrewd categorisation instead of extremely subtle responses to visual clues? No, and John Pearce does not claim that they are. As he says regarding the trees, the birds could be responding to 'a green blob' somewhere in the picture, or to something else we humans might not pick up on. How else do we explain the fact that the birds do quite well judging between the paintings of Monet and Picasso?! 'One way,' Pearce says, 'is to suppose that when the pigeon sees a complex photograph, such as of a Picasso painting, he notices a number of salient features. Then the bird solves the problem simply by looking for those features in the other photographs.'

Pearce does not claim that his pigeons are budding Einsteins. They are capable of surprising performances, but he does not believe they entertain truly abstract thoughts or have an internal dialogue with themselves. They are one step 'below' that level. Nevertheless, he adds, the results on his tests 'suggest we must completely rethink how we view the acquisition of concepts'. He theorises that the tests with the birds and the undergraduates indicate that, while we have two ways of thinking (the concrete and the abstract), the pigeons have only one (the concrete based on the images in their brains). When odd categories like certain abstract shapes happen to be perfectly suited to the pigeons' talents, they beat humans. Otherwise, it's no contest. Furthermore, the pigeons don't even beat goldfish who learn to differentiate between simple shapes and these same shapes with a little bump on the top. Donald Griffin quotes Stephen Walker, a psychologist at the University of London, as being surprised that 'even a vertebrate as small and psychologically insignificant as a goldfish appears to subject visual information to such varied levels of analysis'. But Griffin asks:

> Why should this be surprising, when it is well known that fish can discriminate many types of pattern that signal food or danger? We should be on guard against the feeling that only primates, or only mammals and birds, have the capacity for learning moderately complex discriminations. For the natural life of almost any active animal requires it to discriminate among a wide variety of objects and to decide that some are edible, others dangerous, and so forth.

Alex

We now meet Professor Irene Pepperberg, of the Biology Department at the University of Arizona, Tucson, and her famous African grey parrot, Alex, with whom she and her students have been working for more than 20 years. Alex is unsurpassed among birds in categorisation skills. He identifies – or 'labels', as the experimenters put it – more than 80 objects. He identifies seven colours, five abstract shapes and numerous materials. If Alex is presented with two identical objects and asked to identify the attribute that is different, he correctly answers 'none' in understandable English!

Are his abilities the same as ours when we handle these tests? Pepperberg answers, 'We don't really know. What we would say in a scientific way is that they're analogous.' She estimates Alex's overall skills in these particular tests as equivalent to an ape's or a dolphin's, which are roughly equivalent to a four- or five-year-old child's.

Pepperberg uses elaborate means in her testing to guard against 'Clever Hans errors'. In all the tests, Pepperberg, the official scorer, sits with her back to the testing set-up. She does not know the right answer. She simply reports what Alex has said as an answer. 'For example,' she explains, 'if he says something like "gree", which could be interpreted as either "green" or "three" depending on what you think he should say, a student who knew that the correct answer is "green" might say he was right, whereas I would say it was unclear and the answer would be discounted.' A student trainer who has worked with Alex on a given set of skills would never be the one who tested Alex on that skill. This precaution against 'expectation cueing' keeps Alex from thinking, 'Oh, this is Jo, so we're going to do colour tests today.' Likewise, the bird is always tested in several different ways in the same session, so he can't work out, 'Oh, this is "numbers" today. I'm just going to spout out my numbers.' For example, he might be shown a green key and a blue key and asked to identify the quantity or what quality is the same about the two, what quality is different. He would be asked questions about colour, shape and material.

As with many primates and mammals tested in lab conditions, boredom can be a problem with Alex. It is not unusual for him to say in the middle of a test, 'I'm gonna go away,' and then do just that. But when he's engaged, he's hard to fool. When he says the word for a certain object, he cannot be tricked easily. For example, if Alex says 'wanna banana' and a nut is presented instead, one of three things will happen. He either refuses the nut in stony silence, or he refuses the nut and repeats his request for a banana, or he accepts the nut and throws it at the conniving human being who had the nerve to try to trick him in the first place. The simplest, most straightforward explanation here is that Alex truly does have the banana 'in mind'. It's hard to come up with any other plausible explanation.

One, Two, Three, Four

Many psychologists have long believed that certain cognitive processes are absolutely dependent on language, or on some corollary of language, and that counting is one of these skills. But any number of carefully constructed tests with a range of animals pose a strong challenge to this assumption. In fact, rudimentary counting may be much more widespread in the animal kingdom than we had ever imagined. If this turns out to be the case, even those who most doubt the existence of the conscious animal mind will have to re-examine their position.

The most basic challenge to the idea that counting and language are inextricably linked is posed by the fact that babies, with no language whatsoever, have a rudimentary sense of number, if not of specific numbers. The fascinating textbook experiment here is to set up a little stage in front of a baby and place one toy on the stage. Then hide that toy behind a screen. Next, the experimenter clearly shows another toy also being placed behind the screen. Raise the screen and time how long the baby looks at the two toys before losing interest and looking away. Repeat the test until you have an average time – five seconds, say – that will serve as the baseline for this particular baby. Now introduce a trick into the experiment. Hide one toy behind the screen, place the second toy behind the screen, then raise it to reveal an incorrect number of toys – just one, say, or four. Now measure how long the baby looks at this scene before losing attention. Inevitably, the baby stares at this 'incorrect' number of toys much longer than it stared at the 'correct' number. Something has changed, and babies know it.

So babies without language have some conceptual sense of number. However, babies have the capacity for language, if not language itself as yet. What about primates? It turns out that many primates have exactly the same kind of reaction to this experiment. They look much longer when the screen is raised to reveal an incorrect number of toys. They, too, seem to sense that there's something fishy going on. They, too, seem to have a conceptual sense of number, mass or amount.

On average, a full seven years is required for children to have a solid concept of number. By that age, they can count a small group of jelly beans and name the corresponding number. Hundreds, if not thousands, of experiments have been conducted on a wide variety of animals to determine whether they have a similar grasp of number and counting, and if not, what stage of development they have reached. We should not be surprised to learn that the results of most of these tests are interpreted in many different ways.

In one experiment, which will introduce us to the complexities of the subject, rats are placed in a box containing six identical 'tunnels' lined up in a row. With training, the rats can be taught to locate without fail the fourth, say, of the six tunnels. The specific location of the tunnels within the box is changed from trial to trial; the tunnels themselves are changed daily so that no scent clues have been left behind from the previous trial; and visual clues from outside the box are controlled for. Still, the trained rat will select the fourth tunnel.

Apparently, the rat can 'count' to the fourth tunnel. But is this the concept of 'four' that you and I have? Does the rat extrapolate from knowing about this fourth tunnel to knowing that it has just eaten four pellets of food, or that this is the fourth one of these darned tests it has been given today? The debate rages.

The famous German ethologist Otto Koehler devised a similar experiment which he felt demonstrated basic counting ability – 'thinking in unnamed numbers', in his phrase – by ravens. Koehler trained his birds to select an object based on the number of spots painted on it. It is important to note that these spots were not of uniform size or shape. They could be any size or shape – it didn't matter. Koehler's birds could be taught to unerringly select an object with any number of spots up to seven.

Again, is this 'counting' or something considerably simpler, a merely 'perceptual' understanding akin to our human visualisation of the count of a domino? Shown a domino, we don't really count the six spots on the double-three. We know without counting. Psychologists call this simpler action 'subitizing' and many comparative psychologists believe the animals in most of the classic counting experiments are actually subitizing, not counting. This is the belief of Professor Euan McPhail, psychologist at the University of York, who can be relied on to pour cold water over any assertion of animal consciousness. (McPhail even brings up that old chestnut beloved of early theologians: if animals do have consciousness, do they therefore have souls and, if they have souls, is not heaven a very crowded place, indeed?)

However, subitizing cannot be the explanation for the achievements of certain cormorants in China, whom Dr Miriam Rothschild believes can count to the number seven. Rothschild, who is justifiably famous in natural history circles, reports seeing the birds pull off this feat many times. These particular birds are raised like chickens by fishermen. Fishermen? Yes, because the birds have been helping them with their work for hundreds of years. Cormorants are superb divers for fish, but when they do so on behalf of the fishermen, a string tied around their necks prevents them from swallowing the catch. The bird returns to the surface with the fish in its pouch, and the fisherman takes it out. The unwritten agreement between fisherman and fisherbird is that the bird is allowed to keep for itself every seventh fish. After all, the cormorant, like all of us, needs some kind of incentive for its hard work.

Sometimes the bird dives and returns without a fish, and this failure doesn't count. The bird gets to keep the seventh *fish*, not the fish caught on the seventh outing. How does Rothschild conclude that the bird is counting along with her and the fisherman? Sometimes the fisherman makes a mistake and fails to reward the cormorant after its seventh successful dive, and directs the bird to go down again, with the string still around its neck. When this happens, according to Rothschild, the bird just sits there. It knows it has brought back seven fish, and the last one should be for itself. You say the Chinese translation of 'go on' to the cormorant, but, unless you take the string off, the bird is not going anywhere. Only when the fisherman realises his error and rewards the bird with the seventh fish will the cormorant dive again. After seeing this scenario played out numerous times, Rothschild concluded that the bird was indeed counting to seven.

This is, of course, anecdotal evidence in the extreme, and therefore grounds for dismissal, to many minds. The same cannot be said for Irene Pepperberg's numerical tests with her parrot Alex. Investigating arithmetical ability turns out to be one of Pepperberg's favourite subjects, and these tests are as stringently controlled as the others we have observed. In a typical test of the bird's numerical ability, Pepperberg places in front of Alex a tray of different objects of different materials (wood and wool) and colours (yellow and red). Alex is then asked, for example, how many yellow wooden objects he sees.

The key to this test is the presence of the two criteria – 'yellow' and 'wooden' – which precludes a quick perceptual grasp of the answer. Common sense tells us, and tests confirm, that a perceptual process produces more errors when the number in question is larger, because it's slightly harder to perceptually grasp five objects than two objects. On the 'yellow-wood-objects' test, we humans score the same regardless of the correct answer. This demonstrates that we are *counting*.

The same pattern holds for Alex. He averages about 80 per cent correct answers no matter what the correct answer is. Pepperberg reasons by analogy that this performance suggests that the parrot may also be counting, although she argues that parrots may have better subitizing skills than humans. She has also conducted experiments indicating that Alex transfers his concept of four objects on a tray to four notes played by a piano, but the crucial tests of this ability have yet to be carried out. She is now working with him on associating the spoken word 'four', which he knows to associate with four objects on a tray, and — here's the big catch — the written Arabic numeral '4'. Pepperberg explains, 'Eventually our idea is to present him, say, a green "3" and a blue "4" and ask him to tell us what colour is the bigger number or the smaller number.'

Tests similar to those presented to Alex prove that monkeys clearly understand differences in quantity, if not necessarily the numbers we give to those quantities. The test is simple, as performed by Marc Hauser at Harvard University. In clear view of the test subject, Hauser places two choice plums in one box and one plum in another box nearby; then he allows the monkey to go to one of the boxes and retrieve the food. About 90 per cent of Hauser's monkeys go directly to the box with two plums. Or they go to the box with four plums versus just three. This is a distinction that infant humans do not make until they are three or four years old. After an animal has grasped the basic problem, Hauser adds a complication by mixing food and non-food items and thereby requiring the monkey to keep in mind not only number, but the kind of object as well. Hauser explains, 'In one experiment we would place two plums in one bucket and one plum and one stick or rock in the second bucket. In this situation, again, about 90 per cent of the time they pick the two plums over one plum and one stick or rock.'

These tests differ significantly from Irene Pepperberg's in that they require the animal to keep in mind something it no longer sees. Hauser concludes that the monkeys are counting plums 'in their mind's eye', and he is not surprised by these results: 'Animals are constantly confronted with situations where they must evaluate the number of animals, or the number of pieces of food, so it is a natural task for them to be confronted with solving a simple arithmetical problem.'

Perhaps no primate has been tested in arithmetic more extensively than Sheba, the chimp who works with Professor Sally Boysen at Ohio State University. In one set of these tests, Sheba learns the correct relationship between one or two or three items of a favourite food, and cards with one or two or three black marks. If two jelly beans are presented, Sheba is taught to pick the card with two black marks. This is a simple category test, and Sheba learns it without any problem. Then the test gets harder as the black marks on the cards are replaced by numerals '1', '2' and '3'. Now Sheba is asked to match three jelly beans not with three black marks but with the abstract equivalent, the number '3'. This is a similar test to the one on which Irene Pepperberg is now working with Alex the parrot, although Alex does not see the four items before having to identify the Arabic number '4'. Sheba the chimp masters it easily enough. Next, the jelly beans are dropped from the experiment and Sheba is challenged to master the equivalence between three black marks on a card and the numeral '3'. She picks this up quickly as well.

Now a real counting task is presented to her. She is shown three dishes, each holding zero, one, two or three jelly beans. She is then asked to name the total number of jelly beans she sees. She can do this successfully, up to the number five. Other chimpanzees in a number of different testing situations have also learned to match a small number of items with the correct symbolic numeral. Finally, and most impressively, Sheba's food is replaced by numerals '1', '2' or '3', and she is challenged to add up the total of the abstract numbers by selecting the correct abstract number. And she can accomplish this just as well as she can add the jelly beans.

On the Other Hand

When I want to enter a locked door and don't have a key, I knock. My dogs accomplish the same purpose by scratching on that door with a paw. When I'm hungry I go into the kitchen, open the refrigerator door, and peer inside at the possibilities. My hungry dogs find me, then lead the way to their feeding area of the kitchen. As the question is posed by Stanley Coren in his book *The Intelligence of Dogs* – and, long before that, by Charles Darwin and his protégé, J.G. Romanes – if my pet behaves in the same way as I do in a situation in which I am undoubtedly thinking, isn't my pet therefore thinking as I am thinking? If the chimpanzee Sheba sees five jelly beans and then points to the numeral '5', isn't she therefore counting just as I would be counting?

The answer is . . . maybe, but maybe not, because reasoning by analogy in this way is fraught with danger. It turns out that I owe much of my understanding of this issue to the psychologist Hank Davis, the critic who drew an analogy between Donald Griffin's books and Salman Rushdie's *Satanic Verses*. Davis is not a fan of cognitive ethology, and he explains his objections in his essay in *Anthropomorphism, Anecdotes, and Animals*. Specifically, he informs us of the common logical error known as 'affirming the consequent'.

The following statement is a good example: 'If this is a good car, it will start on the first try.' The sentence is correct, but the implication that starting on the first try proves that this is a good car is quite wrong and could cost me a lot of money. The car could be a lemon and still do this one thing right. With regard to counting, we can say about ourselves, 'If I am counting, I will point to "5".' But it is not true that I point to '5' only if I am counting. There are a host of other reasons why I might do so. Likewise, there might be several reasons why Sheba does so.

It was Hank Davis who conducted the experiment in which the lab rats learned to choose the fourth tunnel, but he does not consider this performance *counting*. He argues that pointing does not prove counting, and he argues that Donald Griffin and other cognitive ethologists regularly make such 'affirming the consequent' errors while assessing the performance of animals on various cognition tests. This is nothing less than the classic error of anthropomorphism, he states. In his view the cognitive ethologists are saying, 'If I, as a human being, am engaging in conscious thought, then I will behave in a sensible manner. Therefore if an animal behaves in a sensible manner, it is engaging in conscious thought.' But the hidden assumption that only conscious thought produces reasonable behaviour is dead wrong. Instinct, for example, produces very sensible behaviour. Davis concludes, 'We do not know enough about the role of conscious thought in determining *human behaviour* [my emphasis] to extrapolate to any other species.'

As it happens – and we should not be surprised to learn this – Davis subscribes to the bold idea that conscious human thought in fact follows rather than precedes action. This is a provocative hypothesis, and there are even some narrowly focused experiments which lend it some credence. That subject is way beyond the scope of this book, but Davis's radical behaviourist leanings don't negate the validity of his warning. Sensible behaviour does not necessarily result from consciousness. The same behaviour across species does not necessarily indicate the same cognitive processes. For example, if I were to get on my hands and knees and start barking along with my dogs, would I be truly barking in the sense that they are? No.

Regarding Sheba's success with numbers, the most popular alternative explanation is that she has simply learned, after arduous trial and error, that when she sees the *shape* of the numeral '2' and the *shape* of the numeral '3', she should pick the *shape* of the numeral '5' if she would like a tasty reward – and she would, of course. This is a plausible explanation, we have to admit.

On the other hand, a fascinating aspect of the tests works against it. When performing counting tests, young children point at objects and count aloud. Chimpanzees cannot count aloud, but they do point and touch and move objects away from the others. Such 'tagging' and 'partitioning' gives every appearance of 'counting' in the same manner as children. Sally Boysen concludes:

> I think we have good evidence that the chimps have a solid understanding of the counting process. We've created a number of situations that require them to come up with novel and innovative ways of putting information together. Now, what this means in terms of how they encode that counting process, or what individual numbers mean to them, is something we can't measure directly, but it looks as though . . . they have an understanding of the number line and something about the process of ordinality . . . very

similar to the capacities we see in children around the age
of three or four.

By the way, Stanley Coren, author of *The Intelligence of Dogs*, is not a
dog lover who believes his pet can handle algebra. He is a professor
of psychology at the University of British Columbia — as well as a dog
lover and prize-winning dog trainer. He tells the following story about
a competing trainer who offers to show Coren the counting ability of
a particular female Labrador retriever, Poco.

'Pick a number from one to five,' this other trainer tells Coren, who
picked three. The trainer proceeded to throw three lures into the high
grass, in different directions, at different distances. The lures were not
visible to the dog, as Coren confirmed by getting down on his hands
and knees to see for himself.

'Poco, fetch,' the man told the dog, with no other apparent cues.
She ran to the lure tossed out last, retrieved it, and returned to her
trainer's side. He accepted the lure, repeated 'Poco, fetch,' and the
Lab headed out and brought back a second lure. Commanded a third
time, she retrieved the third lure. Commanded a fourth time, Coren
writes, 'The dog simply looked at him, barked once, and moved to his
left side, to the usual heel position, and sat down.'

The trainer's conclusion, and Coren's conclusion: Poco knew three
lures had been thrown, and she had recovered three lures. She knew
there weren't any more because she had been counting. Of course,
Coren knows all about the Clever Hans débâcle mentioned earlier.
Could the dog be receiving subtle visual or oral cues, whether
intentionally or inadvertently delivered? Yes, she could be, because
dogs are extremely acute observers, visually and by undetermined
means that might as well be ESP, as far as our own limited senses
can determine. With this in mind, Coren tried in every way he could
imagine to restrict the possibility of cueing, putting the generously
cooperative dog through her counting paces for the next half-hour. He
enlisted the aid of a second bystander who wouldn't know how many
lures Coren had tossed into the grass. This person certainly couldn't
give cues to Poco. In the equivalent situation with Clever Hans, the

poor horse was lost. But Poco nailed this trial, too. The dog always knew how many lures had been thrown and could not be tricked into retrieving a lure that had never been thrown.

But the fact that Poco refused to try to retrieve the lure doesn't prove she had been counting. Hank Davis is right about that. No one can prove that she was counting. But can Davis or the reader come up with a better explanation for this specific set of circumstances?

Somewhat Irrelevant Footnote for Us Dog People

Stanley Coren disproves an idea about dogs put forward by no less a commentator than the esteemed lexicographer Samuel Johnson. 'Did you never observe,' Johnson asked rhetorically, 'that dogs have not the power of comparing? A dog will take a small bit of meat as readily as a large, when both are before them.'

When you think about this statement, it must be one of the silliest that Johnson ever made. Of course dogs compare – if nothing else, the size of other dogs. Is it purely coincidental that a dog is much more likely to move beyond the display stage and actually challenge a smaller dog than a larger one? I don't think so. Dogs know size.

Coren reports a test designed by Daniel Greenberg, the editor at that time of *Science and Government Report*, to refute the great Samuel Johnson. This experiment is easy to set up in your own home. Shape two balls of meat of noticeably different size and place them on the floor directly in front of your dog, one somewhat closer than the other. The dog will almost always go to the nearest ball of meat, regardless of size. Perhaps this was as far as Johnson took his observations. If so, he was not a good scientist. Next, take two more balls of meat of noticeably different size and place them some distance from the dog, say 6 metres (20 feet), but the same distance in each case. Almost invariably, Coren reports and I can confirm (after admittedly limited research) that the dog will run to the larger ball of meat.

chapter 6

TOOLS OF THE TRADE

Not Such an Exclusive Club

If mental mapping is something of an entry-level mental talent for animals, and category formation is one rung up the ladder, the use of tools has long been considered threshold behaviour for true intelligence. In fact, it was widely believed for a very long time that humans were the only tool-using animals. After all, tool use implies a high order of concept formation, problem-solving, adaptability, purposive thinking – the whole range of mental skills cognitive ethologists look for in order to identify intelligence and consciousness. No wonder Charles Darwin's report in *The Descent of Man*, that he had viewed an 'orang' use a stick as a lever, was generally dismissed as another of his anecdotal, anthropomorphic, embarrassing errors that zoologists would rather not talk about.

A hundred years later, Jane Goodall, at her Gombe Research Centre in Tanzania, came up against the same bias when she reported observing a chimpanzee strip a twig of its leaves and stick the resulting stem in a hole with the clear purpose of withdrawing termites. 'When I first saw this,' she says, 'I sent a telegram to Louis Leakey and he became very excited. [But] there was a sort of disbelief among many scientists, because at that time *we* were supposed to be the only tool-using and tool-making animal. Since then we've realised that the chimpanzees use many objects in many different ways and they modify them very, very frequently.'

And not just chimpanzees. Since Goodall's revelation, many books and a large number of academic papers have been written on tool use among a wide range of animals. It turns out that the tool-using club is not nearly as exclusive as we once thought – or perhaps wanted to think. Definitions are disputed, naturally. In the following discussion I will use the strict language employed by the *Oxford Companion to Animal Behaviour*, which defines tool use as 'the use of an external object as a functional extension of the body in attaining an immediate goal'. Fair enough, although we should note for the record that the behaviour, widespread among birds, of dropping a stubborn food item (forgive the anthropomorphising) such as a clam on a rock in order to break it open, then swooping down for the meal, is a clever piece of work, whatever name you give it.

Certain female wasps dig tunnels and then stock them with paralysed insects that will serve as food for the larvae. With the eggs laid and the food stored, the wasp seals the tunnel from the outside. And then some wasps have been observed using a pebble or twig or hardened clod of earth to tamp down the sealed entrance! Is this tool use? Yes it is, and the observations have been made by serious researchers, not just anthropomorphising passersby! Certain ants use little pieces of leaf or wood to soak up fruit pulp or the fluid tissue of prey, and then transport this liquid food back to the colony. Tool use, however you cut it: truly remarkable behaviour.

Of course, these examples bring us into conflict with the old assumption that social insects are purely instinctive creatures. At present, we don't know if such behaviour is strictly instinctive, but let's suppose, for the sake of argument, that it is. How then do we maintain the correlation with intelligence? Perhaps we don't. Euan McPhail and others believe the surprisingly widespread use of tools casts doubt on tool usage as a barometer of intelligence. But why shouldn't there be instinctive tool use as well as purposeful, intelligent tool use? There is no necessary contradiction here, any more than there's a necessary contradiction between the marvellous, instinctive navigation of the arctic tern and the less impressive but still somewhat intelligent navigation of the yachtsman out for a day's sailing around the bay.

When the crows of New Caledonia, the woodpecker finches of the Galapagos, and quite a few other species of birds hold twigs in their beaks to dig for grubs or other small insects, is this tool use? Absolutely. Is it instinctive? To some extent, yes, but birds have also been observed carrying the same preferred twig from site to site. If that's instinct, it's entrepreneurial instinct! One report cited in the *Oxford Companion* tells of a finch who was unsuccessful at fishing for grubs with a forked twig, so it broke the twig above the fork and used the remaining single piece, with better results. Another report describes a young finch raised outside the nest (and, therefore, unable to learn from its parents) who 'manipulated' twigs from an early age, instinctively. When hungry and confronted with food in a hole, this finch dropped the twig and tried to get the food with its beak – unsuccessfully. Then the bird learned to use the twig for this purpose: instinct-plus, I would say.

Perhaps the most remarkable story of tool use and overall problem-solving by birds – by any animal, when you get right down to it – is the incredibly intelligent behaviour of Professor Bernd Heinrich's ravens at the University of Vermont. You almost have to see this performance to believe it. However, the story actually began some years ago in Sweden, where the strangest thing was happening to a group of enthusiastic ice fishermen: someone was stealing their bait and their catch. One Herbert Strandh of Habo wrote a letter to the editor of his local newspaper outlining the mysterious events. Early in the season he got good catches, but after a while his hooks came up empty, devoid of bait or fish. Other ice fishermen reported the same phenomenon. A law-abiding member of a law-abiding nation, Strandh couldn't figure this out, so he set up a stake-out. His letter continued, 'A crow landed near the first hook. Then it walked over to the hole, took hold of the line with its beak and walked backwards until the bait was out and ate it up immediately, whereupon it went on to the next hook.'

The crows were tried and convicted of stealing the bait and the fish! However, perhaps we shouldn't be too surprised by their 'criminal' behaviour. The crow is what ornithologists call a generalist, referring to the bird's feeding habits. This is a category that also includes ravens. Crows and ravens have always had a reputation – in both folk psychology and cognitive ethology – for being highly adaptable and shrewd creatures. Their generalist lifestyle requires such flexibility – and their flexibility allows them to lead such a generalist lifestyle. Reading about the thieves of Habo, Bernd Heinrich decided to find out just how shrewd and adaptable these generalists could be. At the University of Vermont he decided to put the behaviour of the ice-fishing crows to a controlled test, using the common raven as a substitute generalist bird in a classically well-conceived, elegant experiment. A thing of beauty, really.

First, Heinrich raised a group of ravens by hand in an aviary, thereby isolating them from the full range of behaviour they would normally learn in the wild. This assured him that they would come to his experimental set-up absolutely cold, so to speak, with no conceivable direct learning to fall back on when applying themselves to a problem they had never encountered. Then he suspended pieces of meat on string hanging from horizontal poles in the aviary. Heinrich knew what the birds did not know until they tried and failed: the meat could not be snatched off as they flew past. Nor could it be pulled up with one continuous pull on the string. No, this was a difficult challenge for any animal.

'In order to reach the food,' Heinrich writes, 'the birds needed to reach down from a perch, pull up on the string, place a pulled-up loop of string onto their perch, step up with one foot, place this foot onto the pulled-up portion of the string, release the bill from the string to reach down again, pull up on the string, etc., so as to repeat the exact cycle at least five times.'

Then and only then would they win the piece of meat. Five ravens were the first subjects of this test. One solved the problem within six hours. Three of the four other birds finally solved it. One never did. Heinrich concluded what I believe any reasonable observer would conclude: that his birds were able to 'see into the situation' before executing the behaviour, which was simply too complex to be the result of any combination of instincts. He also had the impression –

impossible to confirm – that the four successful birds had not learned this behaviour from each other, but had worked out the problem independently.

Another fascinating aspect of the experiment is what happened when the birds were startled while engaged in the task. In the natural world, a startled raven flies off while carrying with it whatever it happens to have in its beak. Among Heinrich's ravens, the ones who had not yet succeeded at capturing the meat flew off holding the string, only to have it jerked out of their beaks. This would be the predicted behaviour. On the other hand, the birds who had successfully reeled in the meat let it drop before flying off, as if they understood that it was tied to the string and would be jerked from their grasp anyway.

Similar experiments have produced fewer 'successes' than Heinrich observed with his ravens, but no one doubts the validity of his work. Other birds with more specialised feeding habits have never been able to solve the meat-on-a-string problem, but some do show apparently shrewd behaviour. One example that has been written about a great deal is the green-backed heron. Standard operating procedure for this heron and other shore birds is to stand motionless near the water's edge, poised to strike when a fish comes within range. But some shore birds also have an extensive repertoire of techniques to find or attract small fish more actively. The herons have often been seen to toss onto the water a light object that will float – something like a crumb, berry, feather, insect or leaf – and then grab any fish that rises to the bait. Some green-backed herons 'fish' in this manner; many others don't. In some flocks, several birds will pick up on the trick; in others, only one or two birds. So this is not instinctive behaviour, a point made even clearer by the fact that the herons who do fish use different techniques. Some toss the bait from a distance, then swoop down for the meal. Others toss the bait and stand nearby.

We might wonder whether the birds have learned this trick from watching humans, but in fact experimenters have never been successful at teaching herons how to fish with crumbs. The birds apparently pick it up on their own – or they don't. Of course, it is possible to explain these results as classic conditioning, from which perspective the bird learned from a lucky episode in which a dropped crumb was followed immediately by a rising fish. Without the lucky episode, the behaviour would never be learned, and that's why it is so rare. But why discredit the bird because it takes advantage of a lucky accident? Or, as Donald Griffin writes, 'Is it reasonable to suppose that a heron develops this kind of behaviour and uses it successfully without thinking about what it is doing and looking ahead for at least a short time to what it hopes to achieve?'

He doesn't believe so, and neither do I.

To Each His Own

When a sea otter living off the Pacific Coast dives to the seabed and returns with a stone, which it then balances on its chest while floating in the water and uses as an anvil for breaking open a mollusc, is this tool use? Without a doubt. And when the otter then holds this same rock in its 'armpit' while diving for more molluscs, this is *really* tool use — you don't want to lose a good tool once you've found it! This behaviour is not instinctive. Researchers at the Monterey Bay Aquarium have proved that young otters have to be taught, usually by their mothers, to use a rock for cracking shellfish. When an orphaned baby otter is being raised at the Aquarium, it has to learn 'the old rock trick' and many other survival skills from its human 'mother'.

Like the birds, some mammals use tools, others don't. Primates are the chief achievers in this category. According to Vauclair's *Animal Cognition*, at least 17 species are inveterate tool-users. It is part of their daily life in the wild. Many arboreal primates throw branches from trees when pursued (often by human observers, hoping to have branches thrown at them!). Capuchin monkeys are superior tool-users. They spontaneously use branches to drive others away, rocks to break open nuts, and sticks to reach for food. In laboratory tests, they will employ a combination of sticks to reach a peanut placed in the middle of a perspex tube. Bonobos, sometimes called 'pygmy chimpanzees', the fourth and much lesser known of the great apes (the others are the gorilla, orangutan and chimpanzee), are committed tool-users, using leaves to scoop up water, sticks to reach objects or pole vault over a moat, and soft material to wipe their behinds. Frans de Waal writes at length about such tool use in *Bonobo: The Forgotten Ape*.

Still, it comes as no surprise to learn that chimpanzees might be the most proficient of tool-users other than ourselves – after all, they are our closest relatives in the animal kingdom, sharing 98.7 per cent of our genetic material. Chimps and bonobos use tools, make tools, understand tools, and modify the same tool for more than one use, and they use more than one tool for the same purpose. In short, they have what could legitimately be called a tool kit, to use W.C. McGrew's neat phrase. And, to borrow the British psychologist Richard Gregory's acute observation, cited by the philosopher Daniel Dennett, such profligate tool use reminds us that tools not only require intelligence but confer intelligence.

Studies of several groups of Ivory Coast chimpanzees demonstrate both tool use and mental maps in the same situation. The goal for the chimps is the delicious meat of the *Panda oleosa* nut. The *Coula edulis* nut is easier to open and more abundant, but the *Panda oleosa* is prized. However, it is also very difficult to open. (Donald Griffin points out that the loud report which accompanies the cracking of this nut led early explorers to assume that some tribe was forging metal tools.) The chimps do so by using one stone to crush the nut against another stone, or against some other hard surface such as a tree trunk. They also remember where the best tools are, and they use the tool closest to a given tree. In short, they utilise a workshop. It's important to note that, although Panda nuts are common throughout West Africa, only the chimp population of the Tai forest in the Ivory Coast have been observed eating them. As Sue Savage-Rumbaugh writes, in *Kanzi, The Ape at the Brink of the Human Mind*, 'This is a striking example of a *cultural* difference between populations.' That emphasis on 'cultural' is mine. Cultural differences among chimpanzees?! Well, why not?

Rumbaugh continues, 'Even more dramatic, however, was Boesch's observation of mothers actively teaching their offspring the skills of nut cracking.' Boesch is Christophe Boesché, of the University of Basel, who reported the teaching behaviour to Savage-Rumbaugh. She writes:

> Boesch told us how, on one occasion, he saw Salome and her son Satre cracking Panda nuts. Salome cracked most of them, and when Satre tried with a partially opened nut, he placed the nut improperly on the anvil. Before he could strike it with the stone hammer, Salome took the piece of nut in her hand, cleaned the anvil, and replaced the piece carefully in the correct position. Then, with Salome observing him, Satre successfully opened it and ate the second kernel.

On another occasion Boesch saw a different mother instruct her daughter in the correct way to hold the irregular stone hammer. After watching the youth fumble the task for eight minutes, the mother finally came over, accepted the hammer from her daughter, carefully and slowly turned it to the correct position, cracked some nuts and then handed the hammer back.

The next question seems obvious in retrospect, but it was quite bold at the time: could apes learn to *make* stone tools? Could they follow in the footsteps of our cave-dwelling Paleolithic ancestors? (Archaeologists consider the different stone-making traditions as key distinctions between various early proto-human cultures. The first tools, found in South Africa, were made by *Australopithecus* and are dated about one million years ago.) In fact, the apes have not made stone tools in the wild, and have had no reason to, but could they be seduced into trying?

It's a longish story, and well worth the telling. Primatologist Savage-Rumbaugh teamed up with archaeologist Nick Toth at the Language Research Center in Atlanta to find out whether the bonobo Kanzi could be taught to make stone tools. We'll learn a great deal about Kanzi's famed language abilities in Chapter 8, but first we'll focus on the stone tool-making endeavour.

Kanzi had to make a tool good enough – or sharp enough – to

allow him to cut a piece of string securing a box with food inside. He immediately understood what he needed to do and how he needed to do it; he quickly learned to discriminate between good flakes of stone and bad ones; and he mastered – or seemed to master – the action necessary to produce them. (Like all the great apes, bonobos are at least three times as strong as a human of the same height, so the job was well within his physical capability.) But for weeks Kanzi was tentative about making the necessary hard blow.

Then one afternoon, eight weeks into the project, Savage-Rumbaugh was sitting in her office when she suddenly heard 'BANG . . . BANG . . . BANG' and rushed outside to see what crisis had occurred. No crisis at all. There was Kanzi, fracturing rocks with solid blows. Kanzi soon learned to produce sharp stone flakes, but they were on the small side, about 2.5 centimetres (1 inch) long. Savage-Rumbaugh and Nick Toth checked and found that the string was too tough for a flake this small, which would grow too dull before successfully cutting the string. Kanzi needed to make a larger, still-sharper flake.

What happened next surprised everyone, perhaps even Kanzi. Savage-Rumbaugh picks up the narrative in her book:

> One day during the fourth month, I was at the tool site with Kanzi, and he was having only modest success at producing flakes. He turned to me and held out the rocks, as if to say, 'Here, you do it for me.' He did this from time to time, and mostly I would encourage him to try some more, which is what I did that day. He just sat there looking at me, then at the rock in his hand, then at me again, apparently reflecting. I wondered what he was thinking, because he did seem to be pondering weighty matters as he gazed at the rocks. Suddenly he stood up on his legs and, with great deliberation, threw a rock on the tile floor with a tremendous amount of force. The rock shattered, producing a whole shower of flakes. Kanzi vocalized

ecstatically, grabbed one of the sharpest flakes, and headed for the tool site.

That's where the food was tied up with the string. I'm sure there are sceptics who think Kanzi was in fact throwing the rock out of frustration and simply struck lucky. But I trust Savage-Rumbaugh, who concludes, 'There was no question that Kanzi had reasoned through the problem and found a better solution to making flakes. No one had demonstrated the efficacy of throwing. Kanzi had just worked it out for himself.'

Surprised at Kanzi's shrewd success, Savage-Rumbaugh was then surprised that Nick Toth was disappointed, because he was approaching this experiment as an archaeologist. He had hoped to learn from Kanzi's efforts something about the development of the tool-making tradition of early man. Shattering the rocks on the hard floor wouldn't do this investigation any good at all, because that's not what our ancestors had done. So Toth talked his co-experimenter into trying to discourage the throwing technique. Okay, but how? Then another associate came up with the idea of covering the workroom with soft carpet, so that Kanzi would have to go back to 'real' tool-making. But the bonobo had other ideas. After trying a couple of unsuccessful throws against the carpet one day, Kanzi paused, 'looked around until he found a place where two pieces of carpet met, pulled back a piece to reveal the concrete, and hurled the rock'.

As Savage-Rumbaugh writes, 'We have assembled a videotape of the tool-making project, which I show to scientific and more general audiences. Whenever the tape reaches this incident there is always a loud roar of approval as Kanzi – the hero – outwits the humans yet again.'

When spring came, everyone at the research centre moved outdoors, where there was no concrete. Kanzi got progressively better at making stone flakes the crude, old-fashioned way – the human way, that is. But then, once again, he came up with an advance especially suitable for a creature of his enormous strength. One day he placed one rock on the ground, stepped back with another, took careful aim, and fired!

The Famous Dangling Banana

Jane Goodall's favourite tool story from her many years at the Gombe Reserve concerns Mike, who had always seemed to be one of the more intelligent chimps. Goodall says:

> He's the one who learned to use empty paraffin cans to enhance his charging displays and got to be alpha male. [At first, Mike] was a bit nervous of me and I held out a banana to him one day, and he quickly picked a little piece of grass and threatened me. One end of the grass hit the banana, and it was as though a light went on. Suddenly he'd realised he'd got the answer. Instantly he let go of the grass, reached out and picked up a piece of straw, [but] it was obviously 'bendy' and he didn't even try to use it, [so] he dropped it, reached around, found a stick and hit the banana out of my hand. It was an absolute straightforward thought process from the moment the grass touched the banana to the moment he found the appropriate stick.

So it would seem, and so it would also seem regarding 'fishing for termites', but both behaviours have been challenged as indicating not much beyond the minimal adaptation of a natural behaviour. Doubters point out that chimps strip the leaves from branches and vines all the time. They also poke objects into holes for no apparent reason. But the fact that certain 'instinctive' behaviours are utilised in the termite fishing does not preclude the presence of insight. Donald Griffin has made this point many times. J.L. Gould and C.G. Gould write in *The Animal Mind*, '. . . indeed, having the wit to realize that something done in play . . . might solve a novel problem probably does qualify as insight'. These two authors, by the way, are not easily swayed in the debate concerning animal minds. They believe that 'the cognitive underpinnings of a behavior often seem to weaken as we learn more about the natural behavior of a species'. Still, insight is insight, even if it utilises natural behaviours. We might even ask if any insight anywhere, including our own, does not in some way rely on 'natural' behaviour – 'human nature', in our case.

Although the Goulds' statement about insight among chimps pertains to termite fishing, they used it in the context of their discussion of Wolfgang Kohler's famous experiments with chimpanzees in the second decade of this century. Jane Goodall's first report of tool use by chimps in the wild was greeted with scepticism. It turns out that hers was not the first such report, nor the first to be received so dubiously. Wolfgang Kohler encountered such criticism much earlier.

From 1913 to 1920, he was the director of the anthropoid research station of the Prussian Academy of Sciences in the Canary Islands, in the North Atlantic. Actually, he was more or less stranded on those beautiful islands when the First World War broke out in Europe. Kohler subsequently became famous as the founder of Gestalt psychology, but it was in Tenerife that he devised a simple set of experiments that shed a brilliant new light on the animal mind. Every author on this subject discusses Kohler's work at length, greatly aided by the film that Kohler made of his work with these chimps. He also wrote a classic book, *The Mentality of Apes*.

Of course Kohler knew about Edward Thorndike's experiments with cats in the puzzle boxes, and he criticised their value. Those tests had 'hidden' the answer, Kohler pointed out. There was no way the cats or any other animals could have known the correct sequence of mechanisms to push and pull, so therefore they had no choice but to use trial and error. By contrast, Kohler set up situations in which the elements of the answer were visible, just not 'in place'. (It was the different approaches of these two behaviourist-oriented researchers that prompted Bertrand Russell to joke that the animals studied by American behaviourists behave like Americans, 'running about in an almost random fashion', and those studied by Germans behave like Germans, 'sitting and thinking'. Even B.F. Skinner enjoyed the joke and acknowledged the point.)

Kohler's basic set-up in Tenerife was simple. He suspended a ripe banana on a string overhead and out of arm's reach of the subject chimp. In most instances, the chimp in question tried to grab the banana by jumping, grew frustrated, walked off in a sulky mood, stopped and looked back, then looked at some of the items scattered around, then put two and two together and moved back to the scene and used these 'tools' to reach the banana. The 'tools' were merely some wooden boxes and a stick. Some succeeded by stacking up several boxes, but not before the stack had tumbled over a few times. For the *Nature* series our producers reproduced this experiment at Burger's Zoo in Arnhem, Holland. Once again, the boxes proved difficult for the chimps to stack correctly and they took a lot of tumbles. If you set aside your knowledge of the animals' presumed frustration, the footage makes for hilarious viewing. But the chimps do finally succeed.

One of Kohler's chimps, Sultan, became famous in books on animal thinking as one of the first who used a box and a stick to reach his banana. These experiments opened human eyes around the world, and Kohler saw these episodes as evidence that the animals were assessing the situation 'as a whole' (from which we get the term *gestalt*). But the experiments also prompted doubts. No one argued that Kohler's chimps were not clever, but were they as clever as they seemed? Could this be another example of the logical error of affirming the consequent: 'If the ape is thinking, he will use the boxes and the sticks to reach that banana. He does use the box and the sticks, therefore he is thinking.'

Critics immediately pointed out that the chimpanzees never perfect the stacking of boxes. They never seem to catch on to the problems posed, and even after years they stack at random until they hit upon a configuration that will hold them up. Apparently, they are not capable of becoming systematic in their stacking. Other researchers have noted that chimps swing sticks almost automatically in all sorts of circumstances, and put sticks together and stack crates while playing around. So perhaps what was going on here was trial and error, though not of an obvious sort. Clever, maybe, but not what could be called genuine insight.

Coincidentally, exactly the same objection was soon raised against another classic story from the annals of ethology, this one involving blue tits and milk bottles. Blue tits are the relatives of chickadees who became famous in England 'between the wars' for learning to strip off the protective tinfoil tops on home-delivered milk and helping themselves to the rich cream which had risen to the top of the bottle. (It was very rare for milk to be homogenised in those ancient times, of course.) The birds seemed to demonstrate insight, initially, and then imitation, but sceptics approached the blue tit story from a different angle. They pointed out that one natural, instinctive behaviour of this particular bird is to peel the bark of trees in its search for grubs. 'Peel' is the key word here. These sceptics went so far as to set up experiments demonstrating that blue tits will even peel back wallpaper. Time and again. Without reward. Could not the first blue tit have discovered the delicious cream by 'instinctive accident'? And wouldn't other blue tits have followed suit because foraging birds often, if not always, flock to any promising new feeding site? In short, did England's blue tits and Kohler's chimpanzees in the Canary Islands demonstrate insight or something much closer to instinct?

Morgan's Canon proposes that we have to settle for instinct if insight is not clearly required. Scientists can't agree on the answer. To my mind, instinct is a much more reasonable explanation for the blue tit behaviour than for the chimpanzees' behaviour while going after the dangling banana. I repeat the Goulds' perceptive statement in *The Animal Mind*: '. . . indeed, having the wit to realize that something done in play . . . might solve a novel problem probably does qualify as insight'. With the chimps, I would say it definitely does. I would also agree with the Goulds' next remark: '. . . if we take a hard line and require complete, from-the-ground-up novelty before conceding conscious inspiration, then Kohler's pioneering work . . . does not provide much unambiguous evidence one way or the other. But by that standard few humans would get passing marks for cognitive prowess.'

Touché!

I now feel obliged to report the following instructive experiment with humans, cited in the *Oxford Companion to Animal Behaviour* in the entry 'Problem Solving'. Test subjects are provided with two sticks and a string on a nail on the wall. They are then asked to hang a loop on another hook on the wall without getting closer than 2 metres (6½ feet). No problem. Almost all of us will use the string on the nail to tie the two sticks together, then use this device to hang the loop on the hook. Everyone gets an 'A' on this test. But when the experiment is changed and the string is not just hanging on the wall but is now holding up a picture, most of us do not realise that the solution is to convert the use of the string from hanging the picture to tying the two sticks together.

Simple enough, in hindsight, but apparently not in real time. This sobering result is a necessary caution to us when we're thinking about, and perhaps criticising, animal behaviour as a sign of genuine thinking and consciousness. It is easy to debunk such behaviour, but when we stop to think about the overall situation, we realise that a lot of time and practice – years, in many cases – are required by humans to learn new skills, from tying shoelaces to adding 2 and 7, from playing a piano to hitting a ball with a bat. Even as adults we don't find novel situations any easier to fathom and control than animals do, relatively speaking. This simple test proves that point.

Intelligent Enough

Wolfgang Kohler's 'dangling banana' experiments with apes have been replicated to a degree – with pigeons. Robert Epstein and his colleagues at Harvard came up with this clever permutation. First the birds were taught to push a box along a flat surface. Note that they were not taught to hop onto the box, just to push it around. Then they were taught to peck at a banana by standing on a box beneath it. Finally, they were put into a situation with a dangling banana and, off to one side, a box. The birds knew the two skills required – pushing one box; standing on a different box to peck the banana – but would they have the insight to understand that they could use the pushed box to stand on, in order to reach the banana?

In every case, the test bird exhibited some momentary confusion when introduced to the set-up, looking at the banana, then at the box. Epstein continues, 'Then each subject began rather suddenly to push the box in what was clearly the direction of the banana. Each subject sighted the banana as it pushed and readjusted the box as necessary to move it toward the banana. Each subject stopped pushing when the box was in the appropriate place, climbed on the box, and pecked the banana.'

Now the question is: can instinctive behaviour linked with learned behaviour get a poor bird into trouble? The answer is provided by the following experiment, as reported in *Animal Minds*. Pigeons were shown a light *for a few seconds* just before food was presented. The pigeons quickly made a connection between light and food and when the light was again presented the birds instinctively began to peck on it, even though this pecking behaviour was not necessary to receive food. Their instinct is to peck, and peck they do! If a devious experimenter changes the rules of the experiment, so that pecking a shining light stops the delivery of food, it takes a while for the birds to stop their instinctive pecking and eventually learn that a light no longer signals food. In fact, the pecking instinct got them into trouble because they would peck the light even though it *stopped* the food.

Likewise, we have seen how animals use landmarks for mapping purposes, but this reliance on landmarks can backfire. A sooty tern hen, for example, does not so much sit on her egg as sit on the place where she *first* sat on her egg. There's a big difference, which experimenters have proved simply by moving the egg in the tern's absence – moving it just a few inches, in fact. When the hen returns, she sits on the ground where the egg had been, presumably judging this location on the basis of nearby landmarks like rocks. The hen will sit on this exact ground while looking at her egg right in front of her, fully exposed to the dangerous elements.

Like all animals, including us, pigeons and terns give evidence of 'behavioural prejudices', to use a phrase from *The Animal Mind*. However, I was fascinated to learn that pigeons cannot be trained to peck in order to avoid a shock. They can be trained to hop on a treadle in order to avoid this shock, but they cannot be trained to hop on the treadle in order to obtain food. Do these results make sense? Yes, definitely. Pigeons eat by pecking, and they flee danger by jumping or by flying away. Ask them to switch these two instinctive behaviours – to flee by pecking, so to speak, or to obtain food by hopping – and they are in trouble. The same behavioural prejudices hold for lab rats, who can be taught to jump in order to avoid a shock, but who cannot learn to jump in order to obtain food. Jumping for food is not in their innate repertoire, and no amount of learning can compensate for that lack.

A final example of behavioural prejudice involves Sheba, the counting chimpanzee at Ohio State University. In one set of Professor Sally Boysen's tests, she would place on a table two clearly unequal piles of candy-coated chocolate buttons. The rule of this game was that Sheba would receive the pile of sweets she did not point to. So, in order to obtain the greater number of sweets to eat, Sheba should choose the smaller pile. Everything Boysen understood about Sheba's intelligence indicated that this challenge was well within her abilities. Under most conditions, she could understand and master a rule this simple almost immediately. In this case, however, Sheba never learned to point at the smaller pile of sweets. She could not overcome that desire for the bigger pile, even though it was never fulfilled.

(Before we start feeling smug, we should all remember the unsurpassed ability of human beings to engage in counter-productive behaviour. Are examples from either private or public life necessary? I shouldn't think so.)

The bottom line is that every creature's mind, or brain – including our own, of course – is exquisitely tuned by and tuned to the needs of its particular world. There is no contradiction between the fact that a pigeon can perform admirably on certain abstract categorisation tests and the fact that it can then peck away to no avail in a different kind of test. The bird who pecks in vain would never encounter these conditions in the real world, where pecking serves the bird very well indeed – and where, presumably, the bizarre talent that enables it to distinguish Monet's from Picasso's paintings also serves it well.

Rats get along fine without jumping for food. They thrive. Given half a chance, they take over. When interpreting tests of animal minds, it is always necessary to remember the evolutionary and ecological perspectives.

If a seemingly more sophisticated creature does not exhibit the consciousness-implying behaviour of a seemingly less sophisticated creature, does this cast doubt on the importance of that behaviour as an indicator of consciousness? Not at all. What it does is remind us of a vitally important idea that will come up time and again: *all animals alive today are as intelligent as they need to be – and only that intelligent*. (E.O. Wilson, the godfather of sociobiology, proposed the following tantalising corollary. In general, Wilson wrote, the male of the species is less complex than the female, 'in a way that reflects their simpler, more selfish reproductive roles'. Ouch!)

We intuitively understand that the remarkable senses deployed by many creatures are a bonus of evolutionary pressures. For example, cockroaches are about 100,000 times more sensitive to surface vibrations than we are, which makes sense. They need this sensitivity. We don't, thank goodness. Many birds have remarkable eyesight, many predators remarkable senses of smell, and so on. As matters have evolved, they require these finely tuned senses. The same relationship with the environment holds for cognition/intelligence/thinking – use whatever terminology you like. All species have been shaped by the forces of evolution to meet their immediate needs. The more a given species needs to be conscious of, the more it *is* conscious of. Either that or it becomes extinct.

One of the first investigators who thought in these terms was the royal gamekeeper at Versailles in the reign of Louis XV. Charles Georges Leroy published his observations in 1764, focusing on the different environmental pressures on predators – wolves, in this case – and their prey as the most likely explanation for the undoubtedly superior intelligence of the predators. Their 'lifestyle' of cooperation required superior mobilisation and coordination of forces. Meanwhile, their herbivorous prey had only to eat what was there. With this insight, Leroy was a century ahead of his time, maybe more.

But if the use of tools by birds is considered a sign of conscious thought, why do animals that we intuitively feel are 'cleverer' than birds — mammals, with their much larger and more structured brains — very rarely use tools? Even among the primates, only the great apes and capuchin monkeys, to our knowledge, employ tools, and not all of them do so. Others, living under identical circumstances, do not. Is a bird therefore 'cleverer' in some way than an orangutan or a dog? No. It's an 'apples and oranges' question. The vegetarian food eaten by primates has been easily accessible throughout the ages, as far as we know, and their hands are excellent tools in and of themselves. Most of them have never needed other tools. Wolves and dogs don't use tools because as predators (by nature, at least, in the case of dogs) they haven't had to, in order to eat. Other than a weapon or a trap, what device would help a dog or any other predator to catch fast-moving prey? Just as Leroy did with the royal wolves, we see the intelligence of dogs more in their handling of social situations and in their ability to find their way through a neighbourhood.

Since we have certain birds who shrewdly use tools and other birds who are unable to modify maladaptive behaviour, do we therefore have clever birds and stupid birds? Maybe we do, but what we mainly have is birds who know what they need to know. And, as we've seen, every bird — every mammal, every animal, including humans — can be very clever and very stupid from moment to moment, as they (we) encounter situations that they (we) are adept at and then others that are strange or new and therefore create trouble. The reason the classic laboratory animals — pigeons and rats and mice — are laboratory animals in the first place is that they are, in the real world, scavengers and therefore, by definition, versatile, adaptable, creative. Any scavenger should do pretty well on a properly set up laboratory IQ test.

These considerations demonstrate why rating animals on a scale of intelligence requires the judges to make some arbitrary choices. In many ways, some birds are just as intelligent as some mammals, and in every case, it's safe to say, a given animal's highest intelligence will be demonstrated in its natural habitat, where it needs this intelligence. Some animals do much better on certain tests in their natural environment than on the same tests in the laboratory. For example, chimps in the field are excellent at locating hidden food (the chimp rides around on the back of a trainer while another trainer hides food in nooks and crannies). In an indoor laboratory setting, chimps do much worse at this test. The same is true with dogs, who know where they've hidden their bones in the back garden but are not nearly so adept at remembering in a lab setting.

Why can Sarah and other chimpanzees solve the complex problems of analogical reasoning? This is not a trivial question at all. Many researchers consider it a pressing question. Clearly chimps do not need to solve such problems in their natural world in the forest − or do they?! I quote Dorothy Cheney and Robert Seyfarth, from *How Monkeys See the World*: 'Socially living monkeys regularly appear to classify individuals on the basis of kinship or close association, and in many cases they seem based on characteristics other than physical similarity, such as close behavioural association or family membership.'

Cheney and Seyfarth set forth the provocative hypothesis that 'group life has exerted strong selective pressure on the ability of primates to form complex associations, to make transitive inferences, and even to judge causal relations, but primarily when the stimuli are other primates. Monkeys and apes are more likely to solve at least some types of problems if they involve social stimuli, like conspecifics [belonging to the same species], than if they involve objects.'

This hypothesis was first broached by Alison Jolly in 1966 and then elaborated by Nicholas Humphrey. We will hear a great deal more about it later, because it makes particular sense when thinking about the consciousness of our closest relatives, the great apes. These bonobos and chimps, as well as rats and birds, are pretty good at tests in our laboratories, but if you really want to see them or any other animal at their best, you have to observe them in their native habitat. One of playwright David Mamet's characters says, 'The map is not the territory.' Likewise, the laboratory is not life. As the psychologist John Pearce says:

Of course animals are 'intelligent', but that says very little. It's rather like saying all animals breathe. The important questions are how do animals differ in their intelligence and how do they acquire their intelligence in the first place? So when I say to people that I'm studying animal intelligence, what I really mean is that I'm studying the way in which animals learn about the environments in which they live, and I'm studying the way in which animals solve the problems that they confront in their environments.

Grand Prize Winner

This admonition – to think of intelligence in the context of the natural environment – holds particularly true for the beaver. Imagine a beaver in a laboratory – it's a joke. I cannot leave this discussion of tool use and intelligent problem-solving without mentioning what may be the most impressive example of problem-solving in the entire animal kingdom (it's certainly the 'largest' example): the beaver dam. In effect, the dam is this animal's tool, and the building of a beaver dam is an undeniably impressive technical feat, no matter how cold-eyed the observer.

The essential facts about beaver dams and lodges are well known. The first major description of the technology in the annals of American nature writing was by Lewis Morgan, a native of England but on American soil a Renaissance Man of the nineteenth century – elected official, lawyer, anthropologist, businessman and naturalist. Morgan wrote *The American Beaver and His Works*, a staunch defence, in the manner of Darwin and Romanes, of the 'Thinking Principle' in animal behaviour.

Beavers build dams in order to control the water level in the pond that is created behind the dam. The industrious rodents then build their lodge either in the pond or on the bank. In either case this lodge is protected from the high water which would flood it (in the event of a flood, beavers open a hole in the dam), and from low water which might leave the entrance exposed.

Cutting all the trees required for both the dam and the lodge is, obviously, an arduous task, and beavers tackle it ingeniously and efficiently. They clear roads, sometimes installing ramps over immovable obstacles. They may build canals into the forest and float their logs to the construction site. They take advantage of available features such as fallen trees or large rocks. They use braces to shore things up when necessary. They can plug just about any hole. And they build dams hundreds of metres long.

The skills employed and the resulting features are so impressive that you wonder whether beavers can be *that clever*. The answer is probably yes and no. Some of the behaviour required for dam-building is innate. There is no doubt about this. All beavers instinctively saw down trees. They instinctively build little dams. Often, to the human eye, they build excessively – constructing dams that would hold back the Nile. They make mistakes. Sometimes one animal will saw on a tree at one level, his partner at another level. Sometimes a felled tree will catch in a second tree, and beavers apparently do not understand that felling the second tree would then yield a two-for-one benefit. Clever experiments by P.B. Richard have demonstrated that they react to the sound of running water by piling up debris next to a loudspeaker. In fact, even Charles Darwin believed the beaver's sophisticated engineering to be the product of instinct, as he informed Lewis Morgan when the two men met.

On the other hand, beavers do not waste time building a dam if one is not necessary – if they live beside a lake, for example. They do not waste time digging a burrow into a bank that is not steep enough to provide a den above the waterline. If a suitable structure is available in a good location, they won't build a lodge at all. Beavers also seem to employ their basic behaviours in novel ways when necessary, just as we have seen in this chapter with birds and chimpanzees. Donald Griffin, the founder of cognitive ethology, is fascinated by beavers. Griffin cites P.B. Richard regarding what happened in one experiment in which drainpipes were surreptitiously and maliciously inserted into a beaver dam. Initially, the beavers in charge of this dam stuck mud in the dam itself near the drainpipe — that is, near the sound of the running water. Soon, however, they found the opening to the pipe, even though it was some distance upstream from the dam, and plugged it closed.

In another of Richard's mischievous experiments, he placed a pipe whose intake was well upstream and underwater. The entrance to the pipe was covered with a strainer. Eventually the beavers discovered this trick, built a platform directly beneath the opening of the pipe, and then plugged the strainer closed. If any experiment does, this one proves that for beavers, as Griffin writes, 'adding of material to a dam is by no means a rigid response to this particular stimulus'.

One final beaver story, a remarkable episode and a reliable one, reported by my friend of many years, Hope Ryden, in her book *Lily Pond*. Ryden had been observing one family of beavers for some time when one day she returned to the pond to find that vandals — humans — had torn a hole in the dam, which threatened to drain the pond within hours. Ryden and her companion discovered the hole in the morning, before the beavers had arrived for the day's inspection. In the animals' absence, they attempted to help the cause by piling stones in a semicircle around the hole. This structure did slow the outflow, even though most of the stones were below the surface. Mid-afternoon, right on schedule, the chief dam builder arrived at the site to inspect the dam. Immediately he went into action, and was soon joined by three other beavers who had been seldom seen at the dam site.

Somehow they understood that the most efficient immediate course of action was to work on the secondary dam which had been started, unknown to them, by Ryden and her friend. First the beavers tried to fit branches into the stone buttress. When this didn't work at all, they switched tactics and went to the bottom of the pond and the bottom of the wall, plugging the holes in the stone wall there and significantly slowing the outflow. Normally, beavers stick mud and grass onto the upstream side of a dam after they have set in place the infrastructure of branches. In this case, they figured out that working below the surface was more important. It is important to note that by working underwater they were working *away from* the sound of the running water. No simple conditioning response going on here.

The animals worked all night, observed by Ryden. The following afternoon, when they emerged from their lodge, they did something Ryden had never observed before that day: each carried with it a pole from the lodge with which to reinforce the dam. No inspection or 'reinforcement' was necessary. The beavers remembered the crisis situation from the previous night.

The literature on beavers is extensive and replete with such examples of adaptability. As Griffin writes:

> To account for such reversal of their customary behaviour . . . as the thoughtless unfolding of a genetically determined program requires that we postulate special subprograms to cover numerous special situations . . . Such postulation of a genetic subprogram can always be advanced as an explanation of any behaviour that is observed; but the plausibility of such 'ad hocery' fades as their number and intricacy increases. A simpler and more parsimonious explanation may well be that the beaver thinks consciously in simple terms about its situation, and how its behaviour may produce desired changes in its environment . . .

Does the beaver have the dam in mind when it begins its very first labour in that direction? We do not know, but the Goulds write, in *The Animal Mind*, '. . . if beavers do not have a general understanding of the tasks they face, then the power of innate programming extends far, far beyond anything we yet have evidence for'. Too far beyond anything we have evidence for. Innate programming will not explain a beaver dam.

chapter 7

TALKING TURKEY

Alarming News

Donald Griffin, the founder of cognitive ethology, believes that the
case for thinking and consciousness among animals rests on three
pillars: versatile adaptability (which we started to investigate in its
many guises in Chapters 4, 5 and 6); evidence from the neurosciences
(the subject of Chapter 9); and animal communication (the subject of
this and the following chapter). Griffin put the issue very simply in
Animal Thinking:

> As civilized adult thinkers . . . our best – perhaps virtually
> our only – way of learning about other people's thoughts
> and feelings is through verbal and nonverbal communication.
> From these elementary considerations there emerges a very
> simple but potentially potent point . . . If nonhuman animals
> experience conscious thoughts or subjective feelings, we might
> be able to learn about them by intercepting the signals by
> which they communicate these thoughts and feelings to other
> animals. This idea is so basic that scientists, accustomed to
> dealing with complex issues, find it difficult to appreciate its
> potential significance. Nevertheless, the analogy to how we
> learn about other people's thoughts and feelings is so directly
> appropriate that we should make an effort to see where it
> might lead.

In his Harvard office Griffin told me that he began thinking in these terms while arguing about the definition of consciousness with the philosopher Thomas Nagel and other colleagues. Then he added, 'So that's the one idea, it seems to me, that's original and constructive in this whole area.'

That is an overly modest statement, to say the least, but focusing on the importance of communication is certainly a key 'Griffinian' idea. In Charles Darwin's era, every creature's gestures, 'body language' and vocalisations were generally considered to be nothing more than the spontaneous, essentially unconscious expression of emotions – excitement over finding food, or alarm about the sudden appearance of a predator, or an indication, perhaps a warning, of what the signaller was about to do next. In technical terms, this was called the 'indexical' interpretation of animal communication. I much prefer Donald Griffin's facetious label. He called it the 'groans of pain' (or GOP) interpretation.

In the last years of the twentieth century, GOP is no longer tenable. If indeed a vervet monkey, for example, is just expressing an emotional state when it yelps at the approach of a predator, why would it matter whether there is an interested audience for this yelping? The animal's emotional state is presumably the same whether it is alone or with a crowd of other vervets. But the monkey emits alarm cries only if there is another monkey in the vicinity who might benefit from what the signaller 'has to say'. Ethologists label this phenomenon the 'audience effect', and it's a strong clue that there's a great deal more to animal communication than was imagined 100 years ago. Today, the clear empirical evidence, as established by literally thousands of experiments in the field and in the laboratory, is that animals are conveying not only emotion but also information.

What kind of creature intentionally conveys information? A thoughtful creature, Donald Griffin suggests, and many researchers agree with him. But is the lowly chicken, the most 'consumed' bird on the planet, capable of meaningful communication? Could it be possible that a chicken is a thoughtful creature? Let's take a look, because we now know that there is a lot more to farmyard clucking

than we ever suspected. When a cockerel sees an aerial predator such
as a hawk, he issues a different alarm from the one issued after seeing
a fox. Furthermore, a chicken who hears the 'hawk' alarm scurries for
cover while looking up, while the receiver of the 'fox' alarm scans the
immediate vicinity on the ground.

In order to make these findings, researchers equipped cockerels
(golden sebright bantams, to be exact) with small microphones,
because some of the calls are faint. They then rigged up phony
threats, including models of large birds towed overhead on a set
of wires – a contraption which replicated classic experiments by
Niko Tinbergen in the 1950s. In one study, 400 'presentations' of
the overhead hawk elicited 509 alarm calls for an aerial predator, and
zero alarms for a ground-based attack. So there's not much doubt that
the chickens are issuing a specific, meaningful warning, and we have
learned that many other birds also have different calls for different
kinds of danger.

Further experiments will probably prove what experiments have
proved with chickens: the 'audience effect' is manifested in more than
one way. The cockerels in these experiments are more likely to issue
cries of alarm if there is a familiar bird nearby. With only unfamiliar
birds in the vicinity, the cockerels cry in alarm no more than when
they are alone – which is almost never. In effect, the cockerels seem
to be looking after their own.

Chickens also have a 'vocabulary' for conveying information about
food. Apparently they name certain foods, or at least certain qualities
of food, with special cackles. Careful study of sonograms is one proof
of this hypothesis. Another is that nearby birds are more likely to
approach if the indicated food is a preferred one, such as mealworms.
In fact, one report hypothesised that the 'gallinaceous' birds –
pheasants, turkeys and grouse, as well as chickens – have perhaps the
richest diversity of food signalling of any animal. The 'audience effect'
also plays a role here: cockerels call more often if a hen is nearby, much
less if the nearby birds are other cockerels. (On the other hand, the
male birds did not discriminate according to gender with their alarm
calls; with alarms, they are more generous with their information.)

Careful experiments have discovered equivalent system of alarms and responses among squirrels, meerkats and prairie dogs. We probably have no idea how many species have developed systems of differentiated alarm calls. Among the primates, the most widely studied and debated alarm system is that of the aforementioned vervet monkeys, who live all across East Africa. If you have ever sat beside a water hole in East Africa and been mesmerised by the comings and goings and doings of a troop of monkeys with black faces, hands and feet, you've probably been watching vervets. (It's worth remembering that apes, such as chimpanzees, bonobos, gorillas, orangutans and gibbons, are genetically closer to humans than monkeys. Monkeys and apes together are the non-human primates.)

In the 1960s, vervets were reported to have separate calls for aerial predators such as hawks and ground predators such as leopards, and also for a third kind of danger — snakes. (All primates, non-human and human, are terrified of snakes.) These early reports of differentiated warnings raised doubts, not that the different calls elicit different reactions, but whether these different reactions mean that the monkeys have any idea why they are reacting differently. Also, the monkey who issues the alarm almost always takes evasive action itself, so perhaps the other monkeys are imitating the action rather than heeding the alarm cry.

There were a lot of questions, and, in the 1980s, Dorothy Cheney, Robert Seyfarth and Peter Marler went to the Amboseli National Park in Kenya to seek some answers from 11 'social groups' of vervets. Three of the groups were studied continuously for 11 years. Cheney and Seyfarth write about this work in great detail in *How Monkeys See the World*.

An important part of their equipment was a series of hidden loudspeakers, which allowed them to issue bogus alarm calls and to study the resulting vervet behaviour under relatively controlled circumstances. It would take several pages to lay out all the precautions and controls set up by the authors, including blind judging. Today, these measures are standard operating procedure for any scientist working in the field.

To cut a long story very short: Cheney and Seyfarth confirmed that the vervets do indeed use the three reported alarm calls – 'hawk', 'leopard' and 'snake'. They apparently use three others as well, to a much lesser degree – 'minor mammalian predator', 'unfamiliar human' and 'baboon'. ('Minor mammalian' includes lions, cheetahs and hyenas, who are minor in the sense that they rarely attack vervets; baboons will occasionally try to make a meal of a monkey.) The authors confirmed that the monkeys hearing the alarms are responding to the call itself, not to the behaviour of the caller. They also discovered a range of nuances with the 'audience effect'. Like many other animals, solitary vervets do not make an alarm call, no matter how great the danger. Mothers give alarm calls at a higher rate when accompanied by their own offspring than by other offspring. Males give more alarms when accompanied by females than by other males. Intriguing but inconclusive tests suggest that vervets also respond to the predator calls of the superb raven, which, like many birds, has two different alarm calls, one for aerial, one for terrestrial danger.

Critics of vervet communication – those who say other researchers over-interpret these alarms and grunts – make much of the fact that vervets do not adjust their signalling based on the response of their listeners. (The same is true of cockerels in the experiments described above.) The critics argue that this failure of the communicator to modify their behaviour is just one more indication of the limits of animals' language cognition. If animals do have a conscious intent to communicate in a way similar to our own intent, they would be responsive to the effect of that communication, as we are. Because they are not aware, apparently, they are not truly communicating.

Euan McPhail tells us, 'What [animals] do is respond in fixed ways to a fixed set of stimuli, and what the animals detecting those signals do is respond, again, in a very inflexible way – a useful way, for their survival, but very much less flexible than the system we have.' What is really going on, the critics say, is nothing more than classic stimulus-response conditioning, in which the stimulus comes in two parts – a predator and another monkey nearby. It's hard to disprove this. As I've noted in several contexts, a behaviourist can always assert that an 'undetected' or 'misunderstood' stimulus is responsible for any animal behaviour on earth, including our own. (Beware the phantom stimulus!)

Critics also suggest that the vervets' alarm cry means 'Hide in a bush!', rather than 'Hawk in the area!', or 'Run up a tree!' rather than 'Leopard on the prowl!' This might be correct, but so what? Either message is an effective communication of important information. Cheney and Seyfarth take great pains to consider all possibilities. They are not proposing that monkeys have sophisticated language capability. They acknowledge the limits of this communication, and one limit in particular really intrigues them. Vervets employ four basic social grunts. Cheney and Seyfarth ask, 'Why stop at four?' They frequently saw mothers leave their young behind in vulnerable situations. They report one episode in which the left-behind youngster cried loudly for help when a group of baboons approached. The mother turned and saw the situation but could not call out 'follow me'.

That Bird Again

Once again we need some definitions – this time, in order to draw the necessary distinction between communication and language. I take my definitions from Jacques Vauclair's *Animal Cognition* because these are readily understood by us amateurs and they make sense. Communication is the exchange of information between a sender and a receiver using some kind of specific code. Nothing more, nothing less. Language is a system of communication that in addition *represents the world* to a receiver. At first glance the italicised phrase might seem uncomfortably vague, but I believe everyone reading this book grasps the implication. Language is an immeasurably more powerful tool (and it is a tool) for dealing with the world than 'communication' is. The alarm calls and social grunts of the vervet monkeys communicate on the basis of a code, but they do not represent the world in a broader way. They are not language.

For many linguists and behaviourists, this is all we need to know: vervet communication is not language. Cognitive ethologists are happy enough to grant this point, but also feel that this communication tells us a lot about the mind, consciousness and *umwelt* of this monkey.

What about the chatty behaviour of Alex, the grey parrot who has worked all these years with Irene Pepperberg? We have known for centuries that *Psittacus erithacus* and other mimetic birds can mimic our words – they are the only creatures in the world capable of sounding anything like us – and they provide us with a great deal of entertainment in the process; the birds can be taught to say virtually anything, although some sounds are more difficult than others for them to pronounce. However, no one in the cognitive sciences, except perhaps Irene Pepperberg, had any idea that a bird could develop the talent for abstraction that we saw in the preceding chapter with Alex. And he was delivering his answers in spoken English, of course. Is this mere mimicry, informative communication, or true language?

In an effort to find out, Pepperberg worked with Alex in a provocative social context, just as primatologists work with their animals in social settings, but for a somewhat different reason. Pepperberg wanted to keep Alex's attention by capitalising on every parrot's manifest capacity for intense jealousy. He always worked with two trainers, and he was challenged to keep up with the second trainer in earning the attention of the first trainer – Pepperberg herself. She labels this the model/rival (or M/R) technique. One of her colleagues in California is using the M/R model with developmentally delayed children and also with autistic children, and it seems helpful in teaching basic labelling behaviour. Another significant feature of Pepperberg's work with Alex is the complete absence of 'extrinsic rewards', meaning rewards that have nothing to do with what he is learning at the time. Pepperberg feels that such rewards are just confusing. After identifying a cork, say, Alex gets a piece of banana only if he specifically asks for banana. If he asks for apple, he gets apple. If he doesn't ask for anything special, he gets the cork he identified.

Pepperberg acquired Alex from a pet shop near Chicago in 1977, when he was 13 months old. Either she or her students have interacted with him for eight hours of almost every day since then. The work has paid off. Alex has a 100-word vocabulary, divided into basic categories: everyday objects, qualifying adjectives, simple actions such as 'give', and concepts such as 'none', 'same' and 'different'. For 20 years language researchers have been arguing about whether the famous chimpanzee Washoe was making up a new word when he signed 'water/bird'. In those same 20 years, Pepperberg tells us, Alex has coined exactly one new word: 'bannery'. She explains, 'He was being trained on "apple" and he had in his vocabulary "banana", "cherry" and "grape", and instead of learning the label "apple" he insisted on saying "bannery". When you think about it, the apple is kind of like a large cherry and tastes a little bit like a banana. So this linguistic elision, as they call it, made sense.'

Today, Pepperberg is working with Alex on learning phonemes – the basic 'pieces' of sound, such as '*aa*' and '*ee*' and '*th*' that we use to put words together. Can Alex be made aware of how he puts together his words? Pepperberg tells us, 'We put the different bits of sound as refrigerator letters [fridge magnet plastic letters] on a tray and we literally train him to identify the "S" as the sound "*sss*" and the "SH" as the sound "*shh*" and the "N" as "*nnn*".'

When Alex has crossed the 80 per cent threshold in this testing, Pepperberg says, 'We're hoping that he will be able to look at, say, a "K" and an "OR" and another "K" and be able to sound the word "cork".' Pepperberg does not presume that such a performance would be actual reading, but it would indicate that the bird understands 'that his vocalisations are made up of individual bits of sound that can be recombined in novel ways'.

'Does Alex possess language,' Pepperberg asks rhetorically? 'No. Is it complicated two-way communication? Yes.'

Or, as Donald Griffin comments, 'Pepperberg makes no claim that Alex has learned anything approaching the versatility or complexity of human language; but he does seem to have demonstrated some of the basic capabilities that underlie it . . . In short, he gives every evidence of meaning what he says.'

Those Bees Again

I like this statement from Reuven Dukas's introduction to his book *Cognitive Ecology*, in which he quotes an article in *Science* by D.E. Koshland: 'Anyone with a flyswatter knows that the flight information and landing computation of a housefly would elicit the admiration of any flight controller.' Dukas cites the estimate that the fly's tiny brain must cope with a 'visual information load' of about .5 megabytes per second. This assignment is a trivial matter for a 300-megahertz microprocessor, but quite a feat for a fly. We would all agree on that, but are we therefore inclined to credit this winged insect with cognition and consciousness? Most of us would say no, because we tend to think of invertebrates – animals without internal skeletons – as residing on the bottom rungs of the ladder of intelligence, however you want to define that word. Most of us draw the line somewhat 'higher'. But it is not true that 'there's nothing going on upstairs' with these creatures. Insects – by far the great majority of all species, invertebrate or otherwise – exhibit startlingly bizarre but effective and adaptable behaviour. The fly is just the beginning. Exhibit number one must be the famous dance of the honeybee, one of the most amazing and controversial behaviours found anywhere in the animal kingdom.

For centuries beekeepers had realised that their bees buzzed around on the honeycomb in a strange, agitated fashion. They also knew that a particularly nutritious and plentiful food supply of nectar and pollen would attract thousands of bees within a very short period of time: a remarkable, unexplained phenomenon. But no one was more surprised by the explanation that tied together these two observations than the discoverer himself, the Austrian zoologist Karl von Frisch.

Von Frisch had made his mark between the First and Second World

Wars by proving that fish can hear, upsetting the received wisdom of the day. Then he demonstrated that honeybees have colour vision. He was a terrific researcher, but he had studied these insects for many years before realising what had been going on right under his nose, literally, all that time. On the other hand, his failure to see the truth is certainly understandable, because no one had any idea that bees or any other insect (or any other animal, for that matter) could possess such a system of communication. Colour vision, okay, but the second-most complex system of communication on this planet? To the astonishment of linguists, that's how von Frisch described the *Schwanzeltanzen*, or 'waggle dance', in 1946.

As is so often the case in science, serendipity played a major role in the discovery. Alas, the serendipitous event in this case was the Second World War. Prior to the war, von Frisch, a professor in Munich, had studied his honeybees the same way everyone else did, by placing food sources within a few steps of his hives, which he had outfitted with glass windows. It was only during the war, when his Munich laboratory was damaged by Allied bombing and he had to operate from his country estate in the Austrian Tyrol, that von Frisch began to place food hundreds of yards away. Friends and students helped him carry his hives around that splendid countryside. They used – what else?! – bells and cow horns to signal the arrival and departure of a subject bee, identified as such by the little daub of paint von Frisch managed to brush on a hovering insect. In Austria von Frisch realised that the bees' agitated, buzzing behaviour was very different when they were returning to the hive from a long distance than when they had dined just a few feet away. He and his students began a painstaking assessment of this behaviour, and within a few years von Frisch was ready to publish his results and subject them to what he knew would be the harsh scrutiny of the scientific establishment.

Briefly put, von Frisch claimed that the agitated behaviour of foraging bees returning from a significant distance was actually a carefully choreographed dance, described as the shape of a 'figure 8' which has been flattened so that the two sections of the '8' meet not in a single point (as in the type used for this book), but in a line.

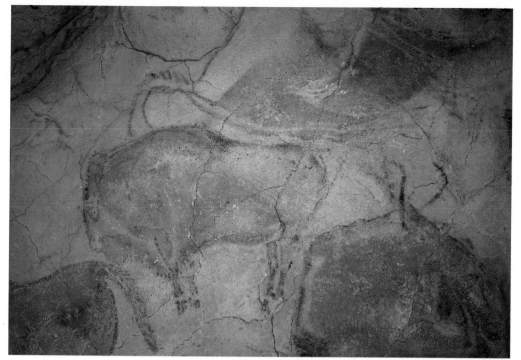

The cave paintings at such places as Altamira in northern Spain and Lascaux in southern France are the oldest surviving examples of visual art. The fact that they depict mainly animals – such as this bison from Altamira – shows how crucial a role they have played in human society from the very beginning.

The famous Clever Hans with his owner, Wilhelm von Osten.

This cartoon from the *London Sketchbook* of 1874 representing Charles Darwin as an ape is part of the huge furore which surrounded his apparent lowering of the status of mankind to the level of an animal. It can equally be said, though, that Darwin raised the status of animals through his idea of a continuum of consciousness from the lowest forms of life to the highest.

Here a rufous hummingbird visits a 'flower' as part of an experiment which proves their considerable capacity for mental mapping.

The Namibian desert is one of the harshest environments on earth. The elephants that live there must cross distances of as much as 64km (40 miles) between their feeding grounds and watering holes. Generation after generation has learned the map of the area and so enabled the population to survive.

In tests where different numbers of plums are hidden from monkeys within their sight, nine out of ten will choose the bucket containing the larger quantity, thus demonstrating an ability to compare quantities.

Dr Nicholas Dodman with his patient Tasher, the neurotic cockatoo.

This woodpecker finch from the Galapágos is using a thorn as a tool for extracting insects from a log.

Even ants are capable of quite complex 'tool use'. These weaver ants are using a larva to 'sew' leaves together.

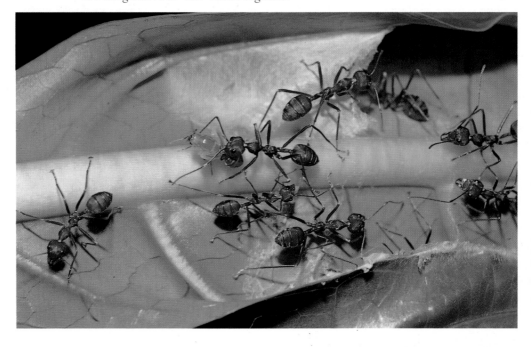

Opposite In the wild, gorillas often sing, especially after a fierce thunderstorm has ended, the sky has cleared and the air smells sweet …

This orangutan at Dr Birrute Galdikas' Camp Leakey Research Centre in Borneo seems to enjoy joining in with the daily laundry. She soaps and rinses clothing enthusiastically, using techniques she has learned purely from observation.

Chimpanzees are perhaps the most successful tool users in the animal kingdom. Here a male uses an improvised nutcracker.

This grass snake is feigning death while remaining totally aware of the approach of a predator. Such techniques, though largely instinctive, must also imply a degree of self-consciousness.

Jane Goodall's work with chimps has been crucial to our understanding of animal behaviour and intelligence.

The question of animal moods is a vexed one. This orangutan seems to be at least bored, though how much of this appearance is our innate desire to anthropomorphise is virtually impossible to say.

You could also think of this flattened '8' as two ovals squashed against one another. Von Frisch labelled this line between the two ovals the 'waggle run'. He said that a returning bee positions his dance so that the angle between this 'waggle run' line and the vertical axis of the honeycomb is the same as the angle between the food source and the position of the sun. If the waggle run is straight up, the food is directly in line with the sun; if the waggle run is straight down, the food is in the opposite direction from the sun; if the waggle run is 45 degrees right of vertical, the food is 45 degrees right of the sun; and so on around the compass. Furthermore, if one bee dances for an extended period in the hive, she compensates for the movement of the sun during that time by shifting the angle of her waggle run! What about the distance to the food source? After decades of further research, the means of designating distance are still not fully understood, but apparently this is achieved by some combination of the duration of the dance and other factors, including perhaps differences in the faint buzzing that accompanies the dance. But distance *is* conveyed, this we know, and every subspecies of honeybee around the world seems to have its own particular formula in this regard.

No wonder the world reacted with astonishment! (I don't exaggerate; von Frisch's report made news headlines everywhere.) This 'dancing' explanation seems impossible, and it turns out to be only part of the story. Foraging bees do not normally dance following their flights outside the hive; they only do so if they have found an especially good source of food. How do they know this? The 'house' bees in charge of unloading the nectar or pollen 'tell' them, in effect, by the speed with which they do their own work.

Among the many bees competing to be unloaded, those with the richest nectar get the fastest service. Those with more ordinary fare to offer the hive have to wait and may even have to find a storage cell on their own. (The house bees judge the quality of the nectar by the residue of the flower which clings to the hairs of the foraging bee. Von Frisch proved that bees are capable of discriminating between one test scent and at least 750 others.)

Any bee unloaded by the house bees within 45 seconds of her

return to the hive will usually dance. Bees who have to wait two minutes almost never dance. But this is not the end of the story, either. The foraging bee also makes an independent assessment of the ease with which a source can be harvested, because only the foraging bee has this information. This assessment includes the distance factor. A good source several thousand yards from the hive is less likely to produce a dance than a lesser source only 100 yards away. What's more, dancers directing other bees to distant sources stop dancing sooner than other foragers as dusk approaches, or during bad weather which would make flying hazardous. Incredible, but true. For their part, potential foragers who are reading the dance also seem to be making decisions. They don't necessarily fly off at the directions of the first dancer they encounter in the hive. And whether the hive needs nectar or pollen or water seems to be a factor in their decision.

I use the word 'seem' because we don't know whether the bees' almost unbelievable behaviour is in fact as goal-oriented as it 'seems' to be. Von Frisch's work was challenged in the years to come by a host of commentators and researchers who suggested that the dance was really just some kind of cooling-off period after a long, hard foraging flight, and that information was conveyed by odour cues. They argued that von Frisch's claims for the directional and distance 'value' of the waggle-run was the figment of an overly active, hopeful, anthropomorphic imagination. Others, including Euan McPhail, are perfectly happy to concede that the dance is a means of conveying information, while denying that there is any conscious intention to do so. By way of provocation, McPhail adds, 'The bees aren't intentionally communicating with each other any more than the flowers that attract the bees with their particular colors are communicating with the bees.'

Setting aside the issue of interpretation, James Gould, one of the authors of *The Animal Mind*, happens to be one of today's leading researchers in this field, and his experiments are among the many that have subsequently proved that von Frisch was correct about the informative nature of the dance. (Von Frisch shared the Nobel Prize in 1973 with Konrad Lorenz and Niko Tinbergen.) Going back to the case of the kidnapped bee, I was especially taken by the experiment in which Gould used a clever scheme to trick foraging honeybees into conveying false information with their dances. They were thus recruiting bees to a feeding site that the forager had never been to. But within minutes of the dance, the targeted site, which had no food and no odour, was nevertheless swarming with bees. There is only one reasonable explanation for this result: these bees had heeded the instructions of the dance.

In another experiment, the Goulds trained a number of bees to fly to a feeding source on a rowing boat in the middle of a lake. Initially, they kept the quality of this nectar low, so that the bees would not dance when they returned to the hive. When the bees were considered thoroughly trained to come to this site, the quality of the nectar was increased and the bees immediately responded in kind by dancing vigorously on their return to the hive. But the key question was: would the other foraging bees follow the directions to the middle of a lake, where there could not be any food, at least not in the 'real world'? Or would they ignore the dance? They ignored the dance – at first, but over a period of time, as the Goulds positioned their boat closer and closer to shore, more and more bees heeded the increasingly plausible directions of the dance.

Totally remarkable.

A robot bee has recently been invented, which the Goulds describe as a 'computer-directed collection of wood, tubing and razor blades'. This bogus bee has had some success in directing real bees to selected sites, opening up the possibility of a whole new set of experiments with very tight controls. It is important to note that the bees' dance is not a perfect system of communication. Recruited bees fly out of the hive in the right direction, plus or minus 15 or 20 degrees, and arrive at the right distance plus or minus 10–15 per cent. Nevertheless, James Gould and Carol Grant Gould estimate that this simple dancing scheme can pinpoint with sufficient accuracy about a billion possible feeding sites outside the hive.

On the other hand, the success of the robotic bee highlights the obvious problem in attributing conscious communication or language to the honeybees. How intelligent can the bees be if they're fooled by this mechanical device? Well, it is certainly correct that the dancing of the bees and the ability to interpret the dance is innate. Bees reared in complete isolation are perfectly good dancers. But, as the Goulds argue (affirming the reasoning of Donald Griffin, who makes the same point in all his writing), 'the presence of an automatic or relatively simple response to a problem does not, in and of itself, prove that a creature is limited to that cognitive level. More challenging circumstances may bring out other strategies that call more of the creature's intellect into play.' They then add, 'We now know that much of the more flexible behavior that appears later in an animal's life is built upon patterns of analysis and response . . . Being able to distinguish between the pragmatic programming success of natural selection and genuinely intelligent behavior . . . is essential to understanding the animal mind . . . however, this distinction is often difficult or impossible to make.'

Donald Griffin brings up yet another tantalising, related question when he writes that the dance and the interpretation of the dance in the hive 'sounds like rather complex behavior for genetically programmed robots'. In recent years, researchers have learned that swarming honeybees also dance to indicate information about a possible new home, communication that is just as complex as the dance for food. It turns out that weaver ants have a system of head wags to direct their fellows to an odour trail which will lead to food sources, and a different set of gestures (when the entire body is jerked backwards and forwards) that indicate intruders in the nest.

If we think about the beehive and the ant bed as they relate to the law of parsimony and Morgan's Canon, we have to acknowledge the role of instinct in the social insects, but we also have to question whether this instinctive behaviour might be too complex to issue from the nervous system in question. These insect species have simple neurological networks. Are there enough neurons, literally, with which to hardwire all the detailed behaviours for Plan A, Plan B, Plan C — all the plans that these social insects exhibit and require? As Griffin points out, the ability to move beyond instinct, to make simple decisions, perhaps generalisations, might be adaptively efficient and beneficial, given a certain point of complexity.

Meanwhile, the sceptics still doubt the capacity of any insect or other animal for symbolic communication – a philosophical position, in effect. As Irene Pepperberg wrote, in her essay in *Cognitive Ethology*, 'By the mid-1970s, it appeared as though "language" was to be defined as that part of human communication that animals were incapable of achieving.' The entomologist E.O. Wilson said this explicitly in *Sociobiology* (the abridged edition): 'With our own unique verbal system as a standard of reference we can define the limits of animal communication in terms of the properties it rarely – or never – displays.' In short, critics of cognitive ethology seem interested only in what animal communication is not: human language.

chapter 8

THE (ALLEGED) GREAT DIVIDE

Two Views

Is there any continuity between animal communication and human language, or are we confronted with an unbridgeable divide? Thinkers throughout the ages have been, well – divided. Aristotle perceived a great deal of continuity up and down the Great Chain of Being. As he put it, in *Historia Animalium*:

> . . . in a number of animals we observe gentleness or fierceness, mildness or cross temper, courage or timidity, fear or confidence, high spirit or low cunning, and, with regard to intelligence, something akin to sagacity. Some of these qualities in man . . . differ only quantitatively . . . for instance, just as in man we find knowledge, wisdom, and sagacity, so in certain animals there exists some other natural potentiality akin to these.

But Aristotle drew the line at language and so arrived at his belief that we are the only rational animals. Descartes also drew the line at language. Darwin disagreed and saw continuity, of course. In *The Descent of Man*, he wrote, 'The difference in mind between man and the higher animals, great as it is, certainly is one of degree and not of kind.' And supporting continuity, Jean Piaget argued that language is a 'particular aspect of the general capacity for representation'. Drawing is another aspect of representation; gesturing is yet another.

Certainly all species with anything like a nervous system operate pretty much the same way at the cellular level. The differences between species are those of organisation — and even organisation is quite similar in closely related species — and sheer brain size. And yet, as Darwin observes, the difference between animal communication and ours is *so* great. Without any question, language is the defining aspect of our human identity. It defines who we are, what we think, and what we do.

Many of us probably think of the chimpanzee in rough-and-ready terms as having a human-like mind, 'minus the spoken language'. But this qualifying phrase may be crucial. Can we really imagine consciousness without language? It is difficult, but is it impossible? How do animals without language do anything that we would consider 'thinking'? We just don't know. Many of us believe they do, but it's impossible to picture the experience precisely, because by this point in our own natural history, everything we think of as part of consciousness has become embedded in language.

Is it possible that within evolutionary continuity 'something happened' that pushed humans across some kind of awesome threshold? Do we humans have so much excess capacity in the frontal lobe that the difference in degree becomes a difference in kind? The stakes are high. As Donald Griffin points out, the only conceivable excuse we could have for denying consciousness to the great apes and the cetaceans is on the basis of language. This 'We have it, they don't' declaration is precisely the position of Noam Chomsky, most famously, but also of many other linguists and philosophers who argue that human language is *sui generis*, the product of a uniquely human 'language organ'. Without language, Chomsky argues, the 'smartest' animal may be capable of impressive accomplishments, but not true thinking. Nor consciousness, suggests Euan McPhail. Animal communication is deemed to be nothing more than a series of possibly sophisticated signals.

I go along with the philosopher Daniel Dennett, who completely agrees with Chomsky that language is indeed the *pièce de résistance* of life on earth, the *sine qua non* of the highest reaches of consciousness.

But, Dennett argues (and the cognitive ethologists agree), the fact that we might not be able to construe the workings of, say, a vervet monkey's brain as 'thoughts', as we understand them, does not necessarily mean the monkey is not thinking. And the fact that the vervet's system of cries and grunts is not human language does not necessarily mean it's not thoughtful communication. The animal may be thinking differently, with what amounts to a different language, in a completely different *umwelt*, and it would be misguided to expect it to answer *our* questions in *our* language.

Ape Language Research

The main battleground for this debate about the ultimate sanctity of human language is ALR – Ape Language Research. It is on this ground that the answers will be revealed in due course. The systematic effort to teach apes some kind of language goes back about 70 years, when A.W. Yerkes, the primatologist for whom the famous Yerkes Regional Primate Center in Atlanta, Georgia, is named, voiced the suspicion that apes had the ability to learn a gestural system. (Actually, the seventeenth-century English diarist Samuel Pepys may have been the very first observer with such hopes. He wrote in his diary entry for 24 August 1661, after seeing an ape in London, 'I do believe it already understands much english; and I am of the mind it might be taught to speak' – about this Pepys was wrong – 'or make signs' – about which he was right.)

In the last few decades, Vicki, Washoe, Nim, Koko, Sarah, Kanzi, Panbanisha and a few others have all become simian celebrities. Vicki was the chimpanzee who learned a few vocal words of English in the early fifties, but that was all, and for good reason: apes' 'phonatory apparatus' (vocal cords, mainly) is poorly adapted for the purpose. When this fact of life became understood, attention shifted to the teaching of American Sign Language or ASL. Washoe and Nim, chimpanzees, and Koko, a gorilla, were all trained in ASL. Then

Sarah, a chimp, was taught to manipulate symbolic plastic chips as her 'words', and Kanzi and Panbanisha, two bonobos, were trained on a keyboard lexigram system.

Today, there is no question that a well-educated ape can learn and employ a large set of sign language gestures or other kinds of symbols, and no question that he or she can tie them together into what are generally called 'strings', in order to distinguish these sequences from grammatical sentences. Such a string might consist of the gestures for 'apple/me/give/quickly'. However, the interpretation of exactly what the apes are 'saying' seems to be in the eye of the beholder. There are believers and there are non-believers. As I read the literature, the belief tends to drive the interpretation, as it so often does in many aspects of life. Those who are anxious, for whatever reason, to believe that the 'language' skill is continuous between apes and humans see something beyond straightforward 'communication' when they interpret this work. They see rudimentary syntax and understanding of the abstract nature of a word. They interpret the string about apples as a genuine grammatical request for an apple. Washoe made news around the world when she spontaneously signed the symbols for 'water' and 'bird' when she first saw a swan; her supporters claimed she had coined a new word in her vocabulary.

At least one trainer made no apology for her anthropomorphic *modus operandus*. Penny Patterson worked with the gorilla Koko in the most homelike setting she could devise and established as close to a parent–child relationship as she could between herself and the female gorilla. Patterson claimed that Koko had an IQ of 85–95 (as measured on the Stanford-Benet test) and a working vocabulary of 500 signs, with another 500 signs used occasionally. Patterson made no attempt to interpret Koko's signing through the cold eyes of the scientist, but rather through the warm eyes of the caregiver. Needless to say, her work was highly controversial within primatology, though there's no denying that Koko did sign a great deal and carry on 'bilingual' conversations in which she responded to spoken phrases with sign language.

Critics of all these projects see intentional or unwitting cueing, as in the infamous Clever Hans episode, or mimicry, or straightforward 'associative labelling'. They point out that many signing 'sentences' may actually mean, to the animal, 'If I do this, then I get an apple.' They point out that almost all signing by the apes is restricted to official training sessions, with not much impromptu use of the symbols; that an inordinate number of the signs and strings consist of requests for food; that any mistake is rationalised as a 'novel meaning'. (For example, Washoe used the sign for 'flower' in reference to smoke. Did this indicate that for her 'flower' meant smells, which is plausible, or that she had no idea what 'flower' meant to her, because it didn't 'mean' anything to her?) The doubters interpret the 'apple string' as not much cleverer than a dog's scratching on a pantry door behind which she has learned her food is stored: clever communication, but not language. And Washoe wasn't making up a new word. She was signing for 'water' and then for 'bird', that's all. And the same goes for other reported neologisms of other apes, such as 'metal/hot' for a cigarette lighter, 'listen/drink' for Alka-Seltzer, and 'candy/drink' for watermelon.

In several sources I read about the following sentence from Nim: 'Give/ orange/me/give/eat/orange/give/me/eat/orange/give/me/you.'

Here, the critics argue, is proof positive of what the apes are really doing: signing anything and everything they know in order to get something to eat, and in the process enticing their trainers into anthropomorphic over-interpretation. Therefore, said one of their more prominent leaders, the linguist Thomas Sebeok, turning the usual understanding on its head, 'Facts do not convince me. Theories do.' In the 1970s, one major researcher, Herbert Terrace, the trainer of Nim (short for Noam Chomsky), put his hands up and announced that he had been mistaken all along. He had now decided that the apes, including Nim, were not learning anything amounting to 'language'. What had seemed in person like thoughtful signings by Nim looked like rote gestures when viewed objectively on videotape.

In an article about the dance of the honeybees James Gould wrote, 'Especially in ethology, it is difficult to avoid the unprofitable extremes of blinding skepticism and crippling romanticism.' This observation is doubly true of Ape Language Research. Looking back at the battles that have raged over the last several decades, Donald Griffin believes that all the criticism was beneficial, because it helped researchers tighten their experimental protocols and controls. Penny Patterson's radical anthropomorphising became even more marginalised, even as researchers came to acknowledge that the whole basis of language training is to anthropomorphise the ape by trying to induce him or her to see and relate to the world as we do, and to 'talk' about it as we do. H. Lyn Miles, a professor of anthropology at the University of Tennessee and a language researcher herself who works with the orangutan Chantek, writes in *Anthropomorphism, Anecdotes, and Animals*, 'How can you raise an ape in a human environment, provide human experiences and encourage human behavior, such as language acquisition or comprehension, to make the ape more humanlike, and then succeed in being nonanthropomorphic in your interpretations?'

Miles is right. It is impossible. As it happens, there has been a great deal of research comparing pre-linguistic communication in human and chimpanzee infants, communication which is wholly dependent on the relationship with the mother. We should not be surprised to learn that there is a world of difference between the two relationships. The mother–child relationship among humans is much more visually orientated than it is in apes, where the contact is more exclusively tactile. There is much less of a teaching relationship between the apes. (Although Washoe's stepson Loulis picked up 50 signs from his stepmother, and only from his stepmother, because no signing was conducted directly with Loulis by humans.) Ape infants interact with human caregivers in a more 'human' way than they react with their mothers – and we *interpret* this interaction in a more 'human' way. How could we not? Miles adds, 'The researcher may not intentionally distort the situation, but the situation itself makes it possible to use more anthropomorphic language when reporting it.' She asserts that the best we can hope for is a 'pragmatic anthropomorphism' – the idea we have encountered earlier, a variant on Griffin's 'critical anthropomorphism'. Some anthropomorphism is inevitable, by definition. We should accept this and allow for it in our interpretations.

As the debate raged on, David Premack concluded that the inherent subjectivity associated with interpreting sign language could be avoided by using symbolic plastic markers of varying size and colour instead. This was how he trained the chimpanzee Sarah, who eventually learned three basic categories of 'words': objects such as 'apple', qualifiers such as 'red', and actions such as 'give'. She also learned a fourth category: 'conditionals' such as 'same', 'different' and 'if-then'. (These are the same basic categories that Alex the parrot was taught by Irene Pepperberg.) If a blue chip stood for a red object, Sarah nevertheless identified the object as red. She understood the symbolism. Sarah knew that half an apple stood in a 'same as' relationship with a half-filled container of liquid. She understood that 'open' could be used in conjunction with a can of food or a box of toys. These tests proved that Sarah had incorporated a level of abstractness in her understanding of what she was doing. This had been difficult if not impossible to demonstrate with the sign language approach, but with Premack's protocols and controls, it was difficult to deny.

Eventually, Sarah understood the difference between these two strings:

'If/Sarah/take/apple/then/Mary/give/Sarah/chocolate'
'If/Sarah/take/banana/then/Mary/no/give/Sarah/chocolate'.

Kanzi

One day many years ago, Sue Savage-Rumbaugh saw a chimp perform a 360-degree flip at the top of his enclosure. Neat trick, she thought, and without really thinking she then made a little loop with her finger. The chimp had never seen this gesture, to her knowledge, yet he immediately understood her request that he do another flip. Savage-Rumbaugh writes, in *Kanzi*, 'I was so taken aback by the alacrity with which he understood what I meant that I simply stood there, rooted in place, staring at him transfixed.'

That moment led to the work that eventually placed the bonobo

Kanzi at the forefront of the ALR debate. First, however, Savage-Rumbaugh had to reconcile her own doubts on the subject. In her early years working with primates at the University of Oklahoma, she became known among her colleagues as 'the unbeliever'. Like Herbert Terrace, she had grave doubts about what was really going on, even though she was impressed by David Premack's work with Sarah and the plastic 'words'. She wondered whether the whole approach of her predecessors might be misguided. These women and men had concentrated on training their charges to produce words. She decided that *comprehension*, not production, was the key cognitive ingredient in language acquisition. But it was precisely the animal's comprehension about which she harboured such doubts. Asked a question such as 'What would you like to do?', most primates in the studies she knew would sign more or less at random.

'Do apes have language?' is the wrong question, Savage-Rumbaugh decided. 'Do apes understand words?' is a better one. For her part, she would focus on understanding what a word meant to her subjects, not on their ability to produce 'strings'. Could an ape talk about 'banana' in some context other than requesting a banana for lunch? As she writes in *Kanzi*, 'It is not obvious that this distinction has been seriously dealt with by other ape-language researchers.' And perhaps the best way of all to find out about true understanding would be to determine whether one ape could tell another ape something it did not know. (Here we recognise Donald Griffin's insight that communication within a species is the best possible window on thought and consciousness.)

In short, Savage-Rumbaugh decided to start fresh with the chimpanzees Austin and Sherman at the Language Research Center at Georgia State University in Atlanta, Georgia. Convinced of the importance of embedding teaching in social situations, as it is with human children, she set out to establish a middle ground between a controlled and an 'enculturated' setting: pragmatic anthropomorphism. And she would use a system of lexigrams, following in the footsteps of David Premack and his use of plastic chips. It is important to understand that these lexigrams are not representative icons. Each shape is completely abstract and arbitrary. Moreover, the location of the lexigrams on the panel was shifted after each work session, to assure that Austin and Sherman were learning the symbols and not their location on the board.

As always, the work was time-consuming and frustrating. Whatever apes are learning, they are not doing so at nearly the rate of children. On average, they learn a few words a month, while children can learn dozens of new words a day. With Sherman and Austin, a pattern set in: frustration followed by the proverbial light going on. The favourite example of most of the commentators I've read is the situation in which one of the two chimps is shown food being deposited by a trainer in one of several containers, while the other chimp is in control of the tools for opening those containers – one tool per container. Without the keyboard, the two passed the correct tool at no better than a chance rate. With the keyboard, would Austin, say, who has seen the food placed in the box, sign to Sherman for the correct tool? Yes. Would Sherman understand the request and pass that tool to Austin? Yes. But, again, this comprehension did not come immediately. When one of the apes was frustrated by the other's failure to get it right, he would gesture through the window, try to get his partner to concentrate on the keyboard, and repeatedly hit the request button on his own keyboard.

After many weeks of training, 'the light went on', and when it did, the two chimps incorporated this use of lexigrams into their daily lives. To all appearances, they were soon playing games with each other via their personal keyboards. This should be no surprise because, as I stated earlier, the use of tools (words are tools *par excellence*) not only identifies but confers intelligence. For the same reason, it shouldn't be any surprise that chimps who have had language training do much better than chimps without this instruction on abstract reasoning problems such as 'key' is to 'lock' as 'screwdriver' is to 'paint can'. Animals with language training understand the 'conservation' test involving glasses of water almost immediately.

In 1980, in the middle of Savage-Rumbaugh's work with Austin and Sherman, B.F. Skinner and associates published what amounted to a parody of her work, featuring two pigeons named Jack and Jill, whom the Skinner team had trained to mimic Austin and Sherman's performance in certain narrow circumstances. By definition, these behaviourists could not countenance the 'intentionality' Savage-Rumbaugh was proposing for her two chimps, but she knew that neither of Skinner's two pigeons would have been able to use the label from the peanut butter jar, spied lying on the floor, to signal the other one about a meal, and to do so spontaneously. That was the level of intentionality and abstraction Austin and Sherman were demonstrating.

That was also the level that Kanzi would surpass. Kanzi is a bonobo, a 'pygmy chimpanzee', which happens to be a very misleading phrase. (Bonobos are about the same size as chimps, but with longer arms and legs; also, they stand upright.) These are by far the least-studied of the great apes, but they may well be the most intelligent. In their native habitat in central Zaire (their *only* native habitat), they are much quicker to pick up nuances of human expression than chimps are. They are much more vocal in the wild than the other great apes. They look at each other and at humans with a very level, considered gaze. As everyone who has ever observed them agrees, it is almost impossible not to anthropomorphise regarding their emotions and behaviour, even more so than with chimps. Frans de Waal writes in his book about bonobos, 'Whatever the implications for language, it is impossible not to be struck by Kanzi's obvious intelligence.' It is impossible to look at the picture of this animal on the cover of Savage-Rumbaugh's two books and not think of him in terms of his intelligence. (*Kanzi: The Ape at the Brink of the Human Mind* was published in 1994. *Apes, Language, and the Human Mind*, a more academic book, co-authored by a philosopher and a linguist, was published in 1998.)

Kanzi was not the first of the great apes to be tutored with a system of lexigrams, but he was the first who used the system spontaneously, after growing up in the company of his adoptive mother Matata, who was learning lexigrams. Without any human prompting, Kanzi became adept at working with the symbols. With Kanzi (and with Austin and Sherman, for that matter), hitting the correct button when shown an apple would not be nearly enough to count as genuine comprehension. They would have to use the lexigram for 'apple' in the absence of an apple, and they would have to use the word to convey information to a third party.

Kanzi also does extremely well with vocalised words – even when heard over earphones, in testing situations, so there is no chance of inadvertent cueing. Not only can he pick up specified objects, he can follow simple spoken directions, such as 'Put the ball in the bowl' or 'Carry the bowl to the refrigerator.' In her book Savage-Rumbaugh details the laborious process she and her associates followed while working with Kanzi. In one study, 13,691 utterances, with a vocabulary of about 100 'words', generated over a five-month period in 1986, were studied individually. After nine years of working with Austin, Sherman and Kanzi with the lexigram keyboards, Savage-Rumbaugh wrote an academic paper for publication. It was turned down by the major journals. (One cited 'over-enthusiastic over interpretation'.) Finally, however, the paper was published in a volume of 'conference papers', and immediately received wide publicity.

Many observers believe that Kanzi and the other apes are indeed exhibiting a 'protogrammar' of limited means, roughly equivalent to the development of a two-and-a-half-year-old infant human, and that they do give solid evidence of being able to communicate simple thoughts, desires and even intentions.

To my mind, the most recent work with apes demonstrates clearly that they are not just 'signalling' in clever ways. They are, of necessity, working with simple ideas. But is this enough? Certainly not for those critics who maintain that nothing short of full-fledged human language counts as any kind of thoughtful or conscious communication. In the effort to defend human 'specialness' – and our language *is* unquestionably special – they denigrate the now well-established and extraordinary abilities that apes do have.

The Real Test is the Real World

Frans de Waal has mixed feelings about language research with the great apes. On the one hand, he writes, 'The apes in these studies are so well attuned to people, so willing to interact, so used to the way we relate to our surroundings, that all sorts of questions can be addressed that are impossible to answer with apes who view us as strangers with strange habits . . . Kanzi's flint-making is an example of an experiment that might not have worked with an untrained subject.'

On the other hand, de Waal has concerns not so much about the anthropomorphism but about the anthropo*centrism* of the whole Ape Language Research enterprise, with its 'Let's see how much like us they are' approach. De Waal believes we might learn more about primates by studying their communication with each other in the wild, not with human trainers in the lab. In *Bonobo: The Forgotten Ape*, he writes, 'Kanzi's wild counterparts wear no earphones, yet they may well listen more carefully to one another than we dare assume.' He wonders whether Kanzi's 'striking command [of spoken English] may actually be better understood in terms of social cognition'. In the wild, bonobos engage in complicated, rapid-fire exchanges, back and back, back and forth, especially during aggressive confrontations – which are relatively rare, by the way, especially when compared with chimpanzees. (As de Waal writes elsewhere, chimps resolve sexual issues with power; bonobos resolve power issues with sex. It is no exaggeration to say that sex, in almost every conceivable form, is the social glue for this species. The authors of the *Kama Sutra* have nothing to teach bonobos.)

De Waal describes these vocal confrontations:

They seem to be trading information about emotions and intentions. The calls are quite variable . . . A spectrographic analysis . . . indicated that they changed in quality over the course of the encounter but remained similar to each other (a process known as vocal matching), as if the two males gradually converged on a solution to their predicament . . . I hold it possible that what chimpanzees do by means of visual displays of strength and determination, bonobos do by means of a more 'languagelike' exchange of information about internal states. It is not that they are talking the issue over, but there is a certain dialectic and coordinated quality to their vocal contests that is absent in the chimpanzee.

Just as the bonobos will never understand our language in all its richness, so we will never completely understand theirs, in all its richness. Cheney and Seyfarth's research with the vervet monkeys uncovered an array of vocalised communication beyond the alarm calls we've already considered. Vervet mothers recognise the vocalisations of their offspring, and they also recognise the vocalisations of the offspring of other specific mothers. A complex set of grunts provides a vital social lubricant among the vervets. Cheney and Seyfarth have pinpointed four of these grunts, each used in narrowly defined situations. They write, 'When a monkey hears a grunt, he is immediately informed of many of the fine details of the social behavior going on, even though he may be out of sight of the vocaliser . . .' Reviewing this research, Donald Griffin writes, 'Although these experiments do not indicate just what meaning the grunts conveyed, they do show that they were not interchangeable.'

Similar experiments have confirmed such purposes among other primates: one kind of cry when meeting family members, another when meeting a highly ranked individual in the troop, and so on. The wedge-capped capuchins of Venezuela combine their calls into pairs, triplets and quadruplets, which seem to reflect 'intermediate' emotional states. It seems quite likely that they make similar discriminations among foods.

Cheney and Seyfarth write, 'The vocal repertoires of nonhuman primates, when assessed by the animals themselves, are far larger than scientists initially perceived them to be . . . Primates make subtle acoustic discriminations when distinguishing between calls.' The key words in that statement are 'when assessed by the animals themselves'. It seems likely that all sorts of communication is being transacted among animals, almost always without our knowledge and understanding. Elephants communicate with infrasound (extremely low sound waves beneath our range of hearing), but we didn't learn this until 1983. We are only beginning to understand the significance of chemical communication with pheromones and other molecules between humans beings. As Donald Griffin said to me, 'Who knows what thoughts and feelings are being communicated and exchanged in the chemical domain?'

The most intriguing story I know of that illustrates the mysteries of animal communication comes from Cheney and Seyfarth, who in turn report one of the studies by Emil Menzel, in which this pioneer introduced one of six young chimpanzees into an outdoor enclosure and showed him either a hidden source of food or a stuffed snake. This chimp, called the leader, is then reunited with his fellows outside the enclosure, and there is no indication whatsoever that the leader has communicated his important knowledge of the situation to the other chimps. Mainly they play and wrestle in the manner of all young chimpanzees. Nevertheless, the moment this gang of six is allowed into the enclosure, they head straight for the food, if this is a 'food' experiment. They know not only that there is food, but where it is, and sometimes the other five chimps precede the leader who has the direct knowledge. If this is a 'snake' test, they all emerge into the enclosure with 'piloerected fur' (a technical term, I guess, but we get the picture) and approach the danger zone with extreme caution, sticks at the ready.

This experiment strikes me as so amazing I'm surprised I haven't read about it more often. According to the school of thought which debunks thoughtful communication among animals, any kind of communication outside that enclosure about specific facts inside the enclosure is probably impossible. But those chimps do communicate, even if we never learn how.

One final, somewhat technical, thought on this issue. Donald Griffin and others have pointed out that English and German are two of the very few human languages in which word order is the key element of syntax, but much of the work in linguistics has been done by native speakers of these two languages. Could this be why such emphasis has been placed on syntactical rules as key elements required of any purported animal language? The vast majority of the thousands of languages around the world (about 7000, but there is no firm consensus on the number) are 'inflected' languages, in which word endings or different word forms are the dominant element in revealing syntactical meaning. ('He', 'his' and 'him', for example, are vestiges of inflected use in English.) Could Kanzi and the other primates be taught even better with some kind of inflected language? Could we be asking our questions in a more effective way?

Dolphins *et al*

Dolphins have brains that are considerably larger than those of the great apes. We should, therefore, expect them to exhibit real talent in the language area – and this is being investigated in Hawaii by Lou Herman and his associates. Herman's bottle-nose dolphins understand the all-important difference between 'Take the Frisbee to the surfboard' and 'Take the surfboard to the Frisbee.' They have no problem with 'right water, left basket, fetch' (which means 'Take the basket on your left to the stream of water on your right') and 'left water, right basket, fetch'.

A charge traditionally levelled at language experiments with all animals has been that almost all the training involves 'imperative' sentences, in which the animal is instructed to do something, or, in the case of the apes, instructs the trainer to do something (provide food, mainly). This complaint was addressed and answered in the work with Kanzi, and it is also addressed in the work with these dolphins, who carry out instructions without complaint, but also answer questions.

'Is there a ball in the tank?' The dolphins answer by pressing one of two paddles, which mean, depending on the experiment, 'yes/present' or 'no/absent'. They have little trouble with these tests. A 'no' answer to the question about the ball proves that the animal both understands the nature of the question and holds in his mind a mental picture of a ball. One day, Herman instructed the dolphin 'hoop, Frisbee, fetch' but there was no hoop in the pool at the time. Herman expected a 'no' response, but the dolphin elaborated on that response, in a way, by fetching the Frisbee and taking it to the 'no' paddle. Herman paused for a moment and than agreed, 'Yeah, that's right.' The dolphins respond successfully to novel sentences. One of Herman's more adept animals, Phoenix, scored at a rate of 71 to 87 per cent on sentences of two or three 'words', and 60 to 68 per cent on sentences with three to five 'words'. (When considering these scores, remember that the chance rate is very much less than 50 per cent, since the dolphins have a large variety of actions to choose from.)

All the marine mammals – including whales, sea lions and otters – have large brains. Those with which such work is practical, including some of the smaller whales, have been the subject of language studies, and all have given evidence of comprehension roughly on a par with the great apes. One of the more successful set of lessons was conducted with the California sea lions Rocky, Bucky and Gertie. After two years, the female Rocky, who received the most intensive training, comprehended 10 signs for objects, five for modifiers of colour and size, and five for actions, such as touching and placing – 20 signs altogether, with which Rocky correctly responded to 190 different combinations.

It has been pointed out that the sea lions and the dolphins could perform well on these tests by following just two basic rules. 'So what?' Donald Griffin asks. Following those two rules might still require genuine thinking. The fact that they and the great apes can follow any syntactical rules at all overthrows centuries of ingrained prejudice that they are not capable of any use of language whatsoever. Griffin writes:

> Two is significantly greater than zero. Furthermore, if dolphins and sea lions think in terms of these rules, they must be capable of thinking in correspondingly complex terms about the relationships between the signals and the actions or objects for which they stand. It would be unwise to allow our preoccupation with the quantitatively unique capabilities of human language to obscure the fact that these experiments reveal at least part of what the dolphins were thinking.

chapter 9

TANTALISING TEASERS FROM
NEUROSCIENCE

Mind or Brain or Both or Neither?

I have no idea what 'mind' really is, and I don't believe anyone else does either, but I do know what the brain is (unlike Aristotle, who believed it was the organ that cooled the body's circulating blood). Somehow thinking and emotion and consciousness are caused, or mediated in some way, by the chemical and electrical activity within the brain and the central nervous system. Or maybe consciousness is *nothing but* this chemical and electrical activity. That could be the answer, according to a lot of modern researchers, for whom Cartesian 'dualism', as it became known, is totally out of the question.

There is a school of thought that mind must be some kind of 'epiphenomenon' which somehow results from the activity of our brains and has no subsequent influence on our behaviour. It is merely a state of the nervous system, an 'emergent property' or an 'intervening variable'. I've read about these three concepts, and I assume that dozens of others are batted around in the ivory towers of academe. All I know is that no philosophy in the world can erase the fact that it definitely feels like something to walk around with this 'epiphenomenon' in my brain.

I wonder whether there isn't a worthwhile analogy to be drawn here with the famous wave/particle duality in particle physics and quantum mechanics. Looked at in one kind of experiment, a photon of light behaves like a wave. Looked at in another, it behaves like a particle. The two explanations seem to be mutually exclusive, but there's no getting around or unifying the experimental results. Physicists have decided to accept this duality as a mysterious fact of life, and to move on to other problems. Couldn't the same perspective be valuable when it comes to the mind–body problem? In *Mind Matters*, Michael Gazzaniga, president of the Cognitive Neuroscience Institute, makes just such a suggestion. He thinks of the mind as the 'interpretive' state and the brain as the 'fluctuating physical-chemical state', both equally real.

In any event, cognitive ethologist Donald Griffin explicitly states that the defence of consciousness for animals does not require the defence of an ineffable, non-physical 'mind'. Brains can be conscious, too; brains can have emotions, too. One of the three main pillars of cognitive ethology, according to Griffin, is the evidence from the neurosciences. So, let's talk about brains and chemicals.

Not a Lot of Differences

I must admit that I didn't understand a great deal of Jeffrey Camhi's textbook *Neuroethology*, but I did remember these two sentences, which I take to be a conclusion based on much of the research reported in the book: 'It is clear that all animals face similar behavioral problems, such as obtaining mates, food, and shelter and avoiding predation. Moreover, the neuron signals in the brain that produce behavior appear to be nearly identical in all animals.' Even at a higher level of structure, the activity in another mammal's brain and in mine is just about the same, when each of us is engaged in similar mental or emotional 'behaviour'. The latest and most detailed experiments using EEGs and ERPs and PET scans and all the other amazing tools are on the way to demonstrating this.

In the *Nature* series about animal minds we observed the high level of activity in the hippocampus of a London cabby as he saw in his mind's eye a particular route through the city; the hippocampus of a lab rat is also the most active part of its brain when it is consulting its own mental map of a radial maze. Experiments show that damage to the hippocampus in any species has harmful effects on whatever spatial memory abilities that particular animal happens to have. For example, lesions in the hippocampus of black-capped chickadees (this organ in seed-storing birds is about twice as large as that of other birds) do not disrupt food storage activity or colour memory, but they do reduce the accuracy of retrieval. The birds' mental maps are damaged. Homing pigeons with analogous lesions initially orient themselves homeward without apparent problems – but then they never get home!

After reviewing the highly technical literature in the field, Donald Griffin writes, 'The behaviorist interpretation of such data is simply that the experiments have monitored electrical activity correlated with information processing, but that this tells us absolutely nothing about . . . conscious thinking.' Nothing is proved, and Griffin doesn't claim it is. His point is that similarities between animal and human brain activity certainly do not lessen the possibility of comparable conscious thinking and emotional life. The neurological similarity is reason enough to take the idea seriously.

Here are two other reasons. In human beings, the left hemisphere is mainly responsible for language; nor is it coincidental that most of us are right-handed, because the right hand is controlled by the left side of the brain. This correlation between the use of tools, the language faculty, and the brain's now-famous 'left-brain, right-brain lateralisation' has been considered uniquely human. It is sometimes cited as proof of the 'We have it, they don't' boast regarding human language. But this assumption of unique lateralisation may well prove to be incorrect. Intrigued by the studies reported in the previous chapter (about the important role that subtle changes in alarm calls, food calls, and other vocalisations play in primate communities), neuroscientists have conducted experiments to determine whether primates also exhibit left- and right-brain biases. These studies do find tentative evidence of lateralisation. Most primates turn out to be right-handed.

REM (rapid eye movement) sleep is implicated in human dreaming, and most if not all warm-blooded animals also have REM sleep. (If invertebrates dream, we will never know by analysis of REM sleep, because their eyes don't move rapidly or in any other way: they are fixed.) Do animals dream? It seems so, based on REM analysis plus the signs of dreaming every dog owner has observed when a sleeping pet starts groaning and churning its legs. And if you have ever mischievously woken an apparently dreaming dog, the results are amusing. The dog reacts as we do when startled out of a deep sleep or dream: a moment is required to work out where it is – if it has not bitten you already! (Let sleeping dogs lie, and all that.) Michael Gazzaniga reports, in *Mind Matters*, on French researcher Michel Jouvet's work with REM sleep in animals. Jouvet concludes that dreams for animals are rehearsals for 'essential, genetically driven response patterns such as . . . fight or flight responses'.

However, there is a rearguard action under way (Euan McPhail is a charter member) which denies that REM sleep necessarily has anything to do with dreaming, because babies give evidence of having just as much REM-like sleep as adults – more, in fact – but what do babies have to dream about? McPhail also notes that animals born at an earlier stage in their maturation – rats without their eyes open, for example – spend almost all of their first days on earth in REM sleep. But are they dreaming? It's an open question, asserts McPhail. I suppose it is, but I can also see how a baby would have a lot to dream about just two hours after its hair-raising entry into this alien world!

Size Matters

The human brain has grown – or, more accurately, evolved – from the bottom up, literally. At the bottom, near the brain stem, are the areas that control emotional life. All mammals have an amygdala and a hippocampus and an olfactory lobe – known collectively as the limbic system. Reptiles and amphibians also have the amygdala and the olfactory lobe. These are the 'oldest' areas of the brain, and they are centres of emotion in all creatures. The amygdala might be thought of as our radar for danger. Activity here triggers fear, the most basic of all the emotions, and stimulates the heart to pump adrenaline. When a hiker suddenly stumbles upon a dozing rattlesnake sunning itself in his path, the amygdala in each animal is chiefly responsible for the fear-based response in each case. Scary pictures alone will bestir the amygdala in humans, as PET scans prove. If Freud was anywhere near correct . . . well, we won't get into Freud, but we can all agree that emotions wield great power in all of us.

On top of these ancient structures evolved the cortex, which controls muscle movement and learning, and the neocortex, seat of the most sophisticated functions of the brain, the highest levels of learning and abstraction. And the neocortex has the important job of mediating the raw emotions pouring out of the limbic system. Only mammals have a neocortex, and comparative anatomy studies demonstrate a direct correlation between the importance of a given sense to an animal and the size of the neocortical area devoted to that sense. In the rat, cat and monkey, for example, significant areas of the neocortex are devoted to sight, hearing, touch and movement. Whereas in the human being, these senses are controlled by a much smaller proportion of the brain. For better or worse, we now have other things on our minds.

Neuroscientists agree that there is a close correlation between brain size relative to body size and the complexity of cognitive activity. When it comes to brains, quantity counts. Beavers, compared to muskrats, for example, have a relatively large brain for a rodent. At the other end of the mammalian scale, there is a threefold difference in brain size between the human being and the chimpanzee. This is roughly the same ratio as exists between the chimp and the hedgehog (about the most primitive mammal there is).

While we humans have trillions of nerve cells in our brains, some worms execute their very basic game plan with fewer than 100 nerve cells altogether. We do not know of a minimum size below which consciousness is 'physically' impossible, but many scientists and laypeople assume there must be some such threshold. The presence of a 'spinal cord' and a big bunch of cells — a brain — at the anterior end are implicitly considered necessary for higher cognitive abilities. This would be bad news for the invertebrates, which do not have a spinal cord, by definition, except that the supposition is simply wrong. As Donald Griffin writes, 'Experimenters have demonstrated that the central nervous systems of crustaceans, insects, and cephalopods organize and modulate information in ways that are quite comparable in complexity and precision to those of vertebrate brains.'

Some invertebrates, like arthropods, a group that includes insects, have a centralised cord of nerve tissue, but it's in the stomach, so to speak, as a system of paired ganglia (bunches of nerve cells). We know that the social insects, in particular, are capable of what appears to be adaptive behaviour. It has been argued that the honeybee's brain (of about 800,000 cells) is simply not big enough to support the cognitive function that would be necessary to support a 'conscious' foraging dance. But Griffin points out that one could argue just the opposite: that their brains are not big enough to support all the complex hardwiring (the ROM, in effect) necessary to program the array of apparently adaptive responses these bees and other social insects are capable of.

Right now, we simply do not know the extent to which brain size constrains a given behaviour, but we do know that the common housefly can perform aerial wonders with its tiny brain. We also know that the 'mushroom body' in the bee's brain — the region generally thought to be the centre of whatever learning the bee accomplishes in its brief lifetime (about three weeks) — gets larger as the bee matures and performs more tasks around and *outside* the hive. I emphasise 'outside' because it is the older bees who do the foraging outside the hive. Is it a ridiculous stretch to wonder whether these foraging bees require increased learning capacity while dealing with their mental maps, foraging choices, and dances?

(The ganglia set-up of the insects even offers certain advantages that we spinal cord-possessors lack. For example, a cockroach participating in an experiment can learn to lift a leg from water in order to avoid a shock even with its head cut off. This handy learning takes place in the ganglia. However, it's only fair to us vertebrates to note that no one, to my knowledge, has ever proved that we, too, could not learn this same trick under the same decapitated circumstance!)

An even bigger problem for the supposition that invertebrates fall below the line of thought or consciousness is posed by cephalopods. Squid and octopuses have brains containing between 100 million and 200 million nerve cells – by far the largest brain among the invertebrates. It is a basic tenet of biology that these cells are doing something worthwhile. Research results are eagerly awaited. Stay tuned.

In earlier chapters we have seen evidence of considerable cognitive powers on the part of birds, even though we tend to think of birds as less intelligent than mammals. These findings on the extent of their cognitive powers are confirmed by anatomy. Comparing brain size relative to body size – the necessary qualification – the largest bird brains are larger than the smallest mammal brains, and there is considerable overlap between avian and mammalian brains all the way 'up the scale' until we reach the primates. It is true that bird brains do not have a cortex of any apparent significance, and since this part of the brain is considered the seat of learning, its absence would not seem to bode well for them, but scientists have determined that it is the striatal areas in bird brains that perform the equivalent functions. Irene Pepperberg, the trainer of Alex the parrot, one of the world's most intellectually impressive birds, has pointed out that the size of the striatal area relative to the rest of the brain varies greatly across avian species. And sure enough, parrots, crows and jays – which are among the smarter birds, however you wish to define 'smart' – have relatively 'hummongous' striatal areas, to use Pepperberg's adjective.

Pet Psychiatry

Nothing has been proved but the evidence is mounting that a brain is a brain is a brain. There are important differences in sheer size, but there also seems to be continuity up and down the line. Regarding the emotions, there can be little doubt. The Cartesian idea of the animal as a mere mechanism without feelings was always ludicrous, and we can now be even more certain of this because of the evidence provided by psychopharmacology. *Medical treatments that work for human beings work for dogs and cats and birds as well.* In my view, this is a profoundly important development.

One of the leading experts in the field is Dr Nicholas Dodman, Professor of Behavioral Psychology at Tufts University School of Veterinary Medicine, and much-in-demand pet psychiatrist. One of the more laid back of the new cognitive psychologists and a charming Englishman, Dodman told me that he acquired his interest in animals from his mother, the 'Dr Dolittle of the neighbourhood' who took care of all the wounded animals. Mrs Dodman felt that the only way to get to know animals was to live as closely with them as possible, when they would act differently and naturally. Her son told me, 'I would go home and she would be standing in the kitchen and some wild bird would swoop through the back door and perch on the kitchen table. She would say, "That's my friend." It was clear this animal did not fear her and they had some kind of relationship.'

Did Dodman inherit or learn his extraordinary empathy for animals? A bit of both, perhaps. He grew up hearing stories about his parents' parrot who had died shortly before he was born. This intelligent bird would curse and bite his father who was pretending to attack his mother. The bird would then ask for an apple. Dodman told me, 'You could only believe that this was a very elaborate series of reflexes or that the animal had actually learned a communication system with people.' By the time he was a student, Dodman had decided that the latter interpretation was the right one.

Earlier in his academic career he wrote papers in which he used a phrase such as 'the animal appeared anxious', and he would be asked what he meant. Well, he said, this creature was vocalising, pacing and hyperventilating in a situation that might well be expected to promote anxiety. Then Dodman's professor would ask how he *knew* this was 'anxiety', because anxiety is a feeling that we could never prove in an animal. In effect, Dodman was being charged with 'affirming the consequent', the logical error we discussed earlier regarding counting. An animal who's anxious will pace and hyperventilate, but this behaviour doesn't prove anxiety. No, it doesn't *prove* anxiety, but can we really imagine any other reason why an animal living in a situation that would promote anxiety in humans would vocalise, pace and hyperventilate just as we would? I don't believe so. Only wilful blindness could deny this correlation. Dodman agrees, but he also had another card to play when dealing with those early critics. He would reply to their doubts, 'Well, how about this: give it an anti-anxiety drug and it's less anxious.'

Can we imagine any reason other than anxiety why an animal which appears anxious should respond to an anti-anxiety drug? Not really, and even Dodman's doubters go along with this. As he points out, they debunk an animal's body language as evidence of an emotional state, but they accept its beneficial response to medicine.

Dodman told me about a paper published in 1992 by a psychiatrist suggesting that dogs as well as humans can have obsessive compulsive disorders. This had never been an issue with Dodman but the paper, by a mainstream psychiatrist at the National Institute of Health in Washington, gave increased credibility to the idea that animals and humans share the same disorders. He told me about the first time he met Klaus Miczek, editor of the *Journal of Psychopharmacology*, and Dr John Ratey, who teaches psychiatry at Harvard Medical School. Independently the three doctors were using the same cocktail of medications for their most aggressive pets, lab animals and human beings, respectively. It was an exciting moment for the pet psychiatrist.

Most of us, in our modern culture, are familiar with the phrase 'alpha male' – the aggressive guy who tries to lord it over everyone else in his quest for dominance. There are plenty of alpha male dogs and cats, and the brain chemistry in each case is practically identical to the analogous chemistry in hard-driving men. One of the chemicals in question is serotonin. Animals – including us – with especially low levels of serotonin are prone to aggression, and they are likely to win the ensuing aggressive encounters. With their dominance reinforced, serotonin levels increase for the time being, all is well, and aggression decreases. When the serotonin levels start to fall again, the likelihood of aggressive behaviour increases. This is a simplified picture of complex events in the brain, but, as Dodman writes in *The Cat Who Cried for Help*, 'it's almost as if dominant animals are serotonin junkies and need to engage in an aggressive encounter to get a fix'.

The cat in question in the book is Ashley, who had manhandled his owner so successfully that he was left at home when the owner went alone to keep the appointment with the pet psychiatrist. But Dodman didn't need to see the cat. The owner's descriptions of biting, nipping, tantrums, and ambushes were quite enough to give a picture of 'a fairly confident and determined cat who liked to get his own way . . .' Dodman handed this owner an 'aggression questionnaire' usually reserved for dogs. Ashley the cat scored quite high. Dodman then suggested a series of behaviour modification tricks for Ashley and the Prozac-like drug Anafranil, which elevates serotonin levels in the brain, thus reducing the impulse to provoke frequent aggressive encounters. 'I have had several aggressive cats become quite placid within a few days of starting Anafranil or Prozac,' he said.

'Well, I feel we should try the medication,' the cat owner replied. 'I just don't feel safe in my house anymore.' The drug worked. The drug almost always works.

Sometimes, however, the cause of aggression is not so much alpha-male behaviour as anxiety. So it was with Stormy, who had been on the verge of driving his housemate Penny crazy with his unprovoked attacks. Of course, Penny had started the problem in the first place by attacking Yoshiko, a new arrival in this particular household of felines. These two cats were sisters, but, as Dodman wryly notes, 'apparently they didn't recognise each other, or if they did, the memories were not fond ones'. Penny despised Yoshiko; Stormy took a liking to Yoshiko. So, predictably, in the resulting triangle Stormy and Penny became sworn enemies. The conflict smouldered for over a year, then escalated when Stormy inflicted serious bodily damage on the smaller Penny, and not once but twice.

The cat's vet had prescribed Valium, which worked well enough for eight days before a dangerous reaction set in. The vet then referred Stormy and his owner to Dr Dodman, who recommended Buspar, 'a designer drug virtually devoid of side effects, toxicity, and addictive potential, and it is an extremely effective treatment for anxiety. It also has anti-aggressive properties'. Stormy calmed down and order was restored to the household.

It turns out that psychopharmacological treatment for cats with emotional and/or severe behavioural problems is perhaps even more common than it is with dogs, because dogs are more 'amenable' to training and retraining than cats are. With cats, medicine is often the only recourse. When the problem is 'inappropriate elimination', as it is euphemistically called (the most common behaviour problem for cats), the solution again may be Buspar, or it may be conditioning, because this is one stress-related disorder amenable to conditioning in cats, or it may even be something simpler, like removing the plastic stair runner from beneath the litter box.(Most cats hate those things, or any other thick plastic.)

One of Dodman's patients was Tasher the cockatoo, an anxious bird who plucks out her chest feathers in a manner that can only be termed an obsessive compulsive disorder, exactly analogous to Trichotillomania in people (in which the individual seeks just the right hair, each one a little bit different, then plucks it and inspects it, and sometimes chews the bulb). Parrots do the same thing with their feathers.

The disorder is not all that rare in psittacines, which are nervous and needy birds at the best of times. Irene Pepperberg's Alex tends to pluck his feathers when Pepperberg is away at a conference. The more she is at home, the better Alex looks. 'That's just the way it is,' Pepperberg says simply, and she interprets feather plucking as 'an artifact of captivity'. In one Dodman study of psittacines all 13 birds with obsessive feather-plucking improved under Fluoxetine, better known as Prozac. The area of the brain that seems to 'control' for such disorders is the striatum, which is, as we have seen, well-developed in birds. And again, serotonin levels are implicated in this disorder.

Prozac works, in Dodman's phrase, like oil on troubled waters. In the end, four of the 13 birds in this study completely 're-feathered'. (Unfortunately, it can be dangerous to medicate birds, given their naturally high metabolic rate; so Dodman gives the average parrot a dose only one-tenth as large as a human's.)

Another of Dodman's patients was Sympathy, a dog who was shot by a policeman two years before arriving at Dodman's clinic. Ever since she had been terrified of policemen, police cars, police dogs, and sirens of any sort. If Sympathy were human she would be diagnosed with post-traumatic shock syndrome. Behavioural therapy would not work with Sympathy, but Prozac helped a great deal. Today, when a police car passes on the streets of Boston, she walks as closely as possible to her owner, but she does not panic.

Dogs (and, to a lesser degree, cats, who are more solitary by nature) are susceptible to separation anxiety and codependence relationships, what Dodman calls the 'Velcro Syndrome'. As with people, this problem results from dysfunctional early experiences. A typical case would be a puppy weaned from its mother too early, then stashed in a cage in a pet shop for weeks or even months. 'When they find a loving owner they bond to them like glue,' Dodman says of such dogs. 'The relationship is more intense than it should be because they're so needy. And the situation gets worse with time, as every separation compounds the one before.' Dodman does not shrink from using the terms 'panic' and 'terror' to describe the feelings of such a dog when the owner leaves the house and the door closes. Serious household destruction is usually the result and people often misinterpret the dog's destructive behaviour as sheer maliciousness and vindictiveness for being left alone. To prove that this is not the case, Dodman suggests that any family in such a situation videotape the dog when it's left alone in the house. What they will see is anxiety and destructive behaviour and the clear evidence of a 'dog who loved too much' (which happens to be the title of one of Dodman's popular books).

'Counter conditioning' helps these animals, beginning with the simple distraction of a tasty titbit when the owner leaves the house. Dodman works closely with the owner to set up an appropriate regimen of 'systematic desensitisation', in which fears are peeled away, one by one. In the case of Max, a golden retriever abandoned by his first owner and painfully attached to the new owner who found him wandering in the woods, this distraction was a bone stuffed with peanut butter. For a moment, at least, the dog experienced a pleasant emotion, the direct opposite of the one he had been feeling at these moments. After a few sessions, Max seemed to forget that he was alone, or even came to enjoy being alone as long as he had a bone filled with peanut butter. Eventually, Max could be left alone with just a few toys, and his howling ceased. This kind of psychotherapy works with dogs and it works with children – a provocative fact, to say the least.

The hardest cases for Dodman to solve are when the unwanted behaviour stems from 'natural' behaviour. The classic case is the dog who chases cars, a reflection of his natural predatory behaviour. The dog locks onto this modern-day prey and loves every moment of the chase, even if he doesn't succeed (and with cars, it is a good thing he doesn't). It's a very difficult habit to break. Many pet disorders, especially eating disorders, stem from the fact that the pet's life, comfortable as it is, is not its 'natural' life. Dodman suspects an element of depression in many overweight house pets. There's also the fact that the animals do not need to burn any calories to obtain food, as they would in the wild, or even outside in the neighbourhood. The answer, for this condition, is a combination of a leaner diet and a richer environment – as part of what Dodman calls the 'Get-a-Life' programme.

Too Much Life

Our pets acquire their emotional problems without our intentional help. However, in the laboratory, cruel experimenters can easily induce all manner of psychiatric disorders in their animals. Intending to test memory, a researcher at UCLA trained a group of rats in a traditional maze, then spiked the food so that the rats became ill, not immediately but hours later. After this much time had elapsed, would the rats put cause and effect together and avoid that food in the future? Yes they would. Furthermore, they acquired a phobia regarding that entire section of the maze, which they wouldn't go near. As Michael Gazzaniga notes, in his entertaining book *Mind Matters*, humans have exactly the same response to a restaurant which has caused food poisoning. We never return to that place.

Other experiments, some of them cruel, demonstrate that animals can easily be driven into a highly stressed state by the random infliction of pain, especially pain about which they can do nothing. In tests with two sets of rats exposed to random shocks, the subjects who could stop the shocks by turning a wheel were better off and developed fewer stomach ulcers than the ones who received identical shocks but couldn't do anything about it. The shocks were the same in each case, but the feeling of being able to do something about it made a big difference in the 'mental health' of the two groups.

Gazzaniga reports studies of large organisations which confirm that the same principle holds for people: senior managers, with more power to control events, feel less stressed than middle managers. Of course, the senior managers also make more money, which might also be a factor in their well-being. When a fish is snared on the hook, its body is flooded with endorphins, the same 'natural painkillers' which flood my own body when I experience pain. Does the fish feel pain? Alas, the inference from biochemistry suggests that the answer may be yes, even though anglers (including myself, most definitely) have always hoped and perhaps believed that this is not the case.

In the wild, certain animals are excellent at inducing stress in each other, with no help from humans at all. Dr Robert Sapolsky of Stanford University has done extensive research with the baboons of East Africa, who have too much spare time on their hands — at least in the case of the alpha male, who successfully makes life miserable for almost everyone else in his troop. He may attack physically, or he may silently harass any other potential 'couple' by simply being nearby at all times. His dominating presence alone is usually enough to make the other male slink off and leave the female alone. The other animals in the group end up with excess amounts of the glucocorticoid hormones – just as humans do when they are suffering from too much stress.

Good Feelings, for a Change

Encountered one on one, sheep are not sheep as we know them by reputation. To know a sheep is to know an animal with quite complex behaviour and good memory, according to Keith Kendrick, a neurobiologist at the University of Cambridge. He knows that a sheep can remember more than 50 members of its flock, that a ewe who has selected a favoured male experiences a surge of dopamine in her brain when shown a picture of that male, and that looking at a picture of a preferred kind of food triggers brain activity in a sheep analogous to the activity in my brain when I look at a picture of a cherry-vanilla ice cream. The seat of that activity in both cases is the hypothalamus, the area of the brain most activated by pleasurable self-stimulation. Do we have any good reason (or any reason at all) to presume that a sheep looking at a picture of choice grass does not feel something like the sense of pleasant anticipation that I do while contemplating the ice cream?

Likewise, grooming among primates and some other animals floods their systems with dopamine. Smiling in humans – and perhaps in other animals as well, if we could read their expressions correctly – is grooming at a distance, in effect. And Dr George Losey of the University of Hawaii has discovered that the little wrasses that clean tiny organisms and other parasites off other fish are allowed to do so only because the action 'tickles' the larger fish. With good tickling, the fish will stay parked for hours, but with bad stimulation they will bite the offending cleaner fish (or the offending fake fish, if we're talking about one of Losey's experiments). One fish practically lifted itself out of the water in an effort to rub against a piece of wire just above the surface. As far as Losey is concerned, 'The sea is not full of cold fish at all. They're out there having a darned good time. They get tickled during the day, they get sex in the afternoon, everything is probably a pretty place, they have a lot of pleasure.'

No one can say that emotions are chemical states, and nothing but these states, in the brain; but there's no doubt that they are mediated by the brain's chemical states in some way. Lab rats can be seduced into pressing a lever almost endlessly if the action results in either the direct injection of endorphins, such as dopamine, into the brain, or the electrical stimulation of the hypothalamus, which then releases the dopamine. Dodman believes this is strong evidence that emotions in animals are essentially the same as the emotions we feel. Or, to put it another way: wherever it's found in the animal kingdom, dopamine is dopamine.

And oxytocin is oxytocin, as we learn from Dr C. Sue Carter, at the University of Maryland, who conducted an ingenious study of the prairie vole. Oxytocin is the hormone that seems to play a key role in the experience of pleasure we feel during sexual activity. It is released in the brain of all mammals at these pleasurable junctures. It also plays a key role in mother–infant bonding (indicated by its presence at elevated levels at the time of birth and when nursing). Oxytocin is a 'relaxing', anti-stress hormone and, apparently, much more. Thinking about these matters, Carter hypothesised that the hormone might also be instrumental 'earlier' in the behavioural cycle that eventually leads to sexual activity and child-bearing. The term 'emotional glue', sometimes used in reference to human couples, doesn't mean much to biologists, as Carter points out. She wanted to investigate the biochemical underpinnings of this 'glue'. Specifically, she wanted to find out whether the same biochemical glue that underpins sex and maternal bonding in people and voles – oxytocin – also underpins pair bonding.

The prairie vole is a good subject for studying this hypothesis because it happens to be a socially monogamous species. (Such behaviour is quite rare among rodents. I should also point out that social monogamy is not the same thing as sexual monogamy. For example, marriage among humans is an institution of social monogamy, but it does not guarantee sexual monogamy – adultery happens. The same holds true for these voles. But promiscuity doesn't matter to Carter, who is interested in the initial and permanent pair-bonding behaviour.)

At her laboratory in Maryland, Carter introduced a young, unpaired female vole to a male and allowed her to spend one hour with him. Under normal circumstances in the wild, this one hour is not enough time for the female to 'make up her mind' about this potential lifetime mate. In the wild, she'll go back and forth between candidates, vacillating, undecided, for about 24 hours. Her decision is indicated when she begins spending almost all her time with one of the males, and may even become aggressive with the 'loser'. In effect, female prairie voles go through puberty in a 24-hour period. Prior to meeting their first strange male, their reproductive hormones are suppressed.

So one hour is not nearly long enough, unless the pre-pubescent female is injected with oxytocin prior to meeting the male – any male. In this case, bonding and mating will almost certainly follow within the one-hour time constraint. Carter concludes that the hormone biases her in favour of whichever male she meets immediately following the injection.

Clearly, oxytocin precipitates bonding, and Carter's work also confirmed that its absence prevents bonding. If the natural release of oxytocin in the female was blocked by biochemical means, she would not bond at all, no matter how long she was kept in contact with a male. This one hormone cannot be the whole story, Dr Carter says, but it must be a major part of that story.

She is careful to note that oxytocin 'works' only if the hardware (the neural receptors) are present upon which oxytocin can act. If you give this hormone to a rat, for example, this rodent does become somewhat more social, but it does not become monogamous like the prairie vole. Give the hormone to a different kind of vole that is not monogamous, and this vole does not suddenly become monogamous, either.

Carter concludes:

When a young prairie vole male meets a female, they go through a very dramatic hormonal shift which might have some of the same features as a human's 'falling in love'. We don't know, we don't have measurements of 'love' in prairie voles, we only have measurements of pair bonds and attachments . . . It is particularly difficult to study 'positive' emotions. Most of the research on [animal] emotion deals with 'negative' emotions such as fear and anxiety. The work we're doing with prairie voles is a bit unusual. It allows us to focus on a positive emotional experience that animals have.

It also provides evidence that even the lowly prairie vole may be experiencing emotions, not totally unrelated to our own, at crucial moments in its life. Could the prairie vole actually 'fall in love'? Apparently some people find this possibility alarming and 'dehumanising'. I don't. I have no problem giving any animal its due.

chapter 10

THE EMOTIONAL LIFE OF ANIMALS

Can There Be Any Doubt?

Hormones are not emotions. Even behaviour is not emotion. Only emotion is emotion. The biochemical evidence for animal emotions is compelling, but it is not ironclad proof that animals feel emotions as we do. When we speak of anger we do not have a shortage of serotonin in mind; when we humans dream of love we are not imagining a dose of oxytocin coursing through our brains. We mean the subjective feeling of anger, the feeling of love. When we have those feelings, biochemistry and 'behaviour' are utterly beside the point.

Dr Marian Stamp Dawkins, Oxford zoologist and author of *Through Our Eyes Only? The Search for Animal Consciousness*, has come to believe that the weight of evidence is on the side of consciousness for animals, but she admits that we cannot prove animals have these subjective feelings, and she warns against our natural tendency to accept at face value the body language and behaviour we do see. The dolphin's 'smiling' visage just happens to be the way its head is shaped. The chimpanzee's biggest 'grin' is actually an expression of alarm. (Editors in the era of the early manned space programme may not have known this when they featured cover photos of 'grinning' chimps emerging from their capsules after an orbit or two.) And is a charging lion 'angry', or just charging? 'Just charging' is more likely. The lion's only object in attacking the wildebeest is eating. It's not personal. We humans can certainly attack one another without anger.

Millions of soldiers throughout the ages have been killed by millions of other soldiers who were not angry. When we think of one-on-one aggression, anger usually plays a role, but not always. One sibling can strike another not out of anger but as a way of trying to establish dominance.

So we do have to be careful about equating hormones, body language and even behaviour with emotions. On the other hand, we have plenty of other clues that the correlation is highly likely. One is that animals incontrovertibly *anticipate*. This might not seem like an important point, but it is. The early experiments proving that animals experience anticipation helped undermine the strict behaviourist agenda, which had little room for such a 'phenomenon'. Rats were supposed to hustle through the maze and accept their food at the end. If they were tricked and there was no food, well, there was no food. The rat was not expected to act surprised when the reward turned out to be missing. But a rat does act surprised in this situation, and it was difficult for behaviourists to explain, because such surprise implied not only memory, which was acceptable as a factor in conditioning, but also anticipation and disappointment, which were not. (The first experiments in this regard were conducted by Edward Tolman, whom we met in Chapter 4, the early twentieth-century investigator who had doubts about behaviourism all along, but felt constrained by the dogma until later in his life.)

Marian Dawkins writes about a 'remarkable' series of experiments with humans and rats conducted by Michel Cabanac, a Canadian physiologist. 'His experiments', Dawkins writes, 'show that the physiological and behavioural [responses] are so similar that the leap of analogy we would have to make to assume that conscious experiences are also similar is reduced to the barest minimum.'

In the experiments, Cabanac charted the reaction of humans and rats to sweet substances. Humans graded a series of sugary solutions on a scale running from '+2' for 'really good' to '−2' for 'really bad'. Then he plotted changing attitudes towards the same solution depending on circumstances, such as hunger, thirst or a prior sweet drink. And these attitudes did change, as we would expect. Then the rats took the same test. Since they couldn't report on their opinions, Cabanac judged their reaction based on how much they drank of each solution under the different circumstances. Plotted on a graph, the results were astonishing: the subjective experience of pleasure/no pleasure reported by the human subjects matched almost exactly the quantity of the solutions drunk by the rats. Dawkins argues that this is a powerful justification for drawing an analogy between the conscious feelings of pleasure experienced by both groups of subjects. Of course it is possible, as Dawkins states, that we with our conscious feelings just happen to react in the same way as rats who are nothing but little machines. But is it plausible?

Pain is not an emotion, but it is another clue to emotion. Strictly speaking, pain is a warning that what's going on in the immediate world right now is dangerous to good health; we should take immediate action to alleviate the condition causing the pain. Given this obvious and undisputed function of pain for *Homo sapiens*, it strikes me as absolutely remarkable that Descartes, and many others with much more technical knowledge than he had, could argue that animals do not feel pain. If pain is not felt, then it's not pain and it doesn't serve its purpose. But it does serve its purpose. When I accidentally step on my dog's tail or her foot, she howls bloody murder. What could provoke those howls if not pain? So it seems to me that we are perfectly justified in concluding that pain is, therefore, very much felt by my dog.

If this logic isn't totally scientifically airtight, too bad. (Or – with apologies to Monsieur Descartes – *quel dommage.*) From my layman's point of view, common sense is sometimes quite enough – thank you very much! It is also true that the experience of pain is mediated by culture. For example, rhesus monkeys inflict severe wounds on each other when fighting, and respond to those wounds as we would, by screaming. But these outward signs of distress very quickly disappear and the animal is utterly quiet. In the mating season, one rhesus may go so far as to rip out an opponent's testicle. The wounded party will scream bloody murder for a while but then, remarkably, he will be seen mating later the same day. Many observers have drawn a similar conclusion regarding childbirth, which has evolved into something of a long ordeal in Western culture, but only a day's – or an hour's – absence from the fields in many other cultures.

We know that animals can carry pain in the form of 'emotional' scars for a lifetime. This is the downside of the extraordinary effectiveness of the amygdala at inducing fear and prompting the necessary evasive action. Take Keiko, the orca, or 'killer whale', who starred in the film *Free Willy*. In September 1998, Keiko was released into a 'sea pen' off the coast of Iceland after living – surviving, I should say – for almost two decades in a small oceanarium in Iceland, another in Canada, a dinky local amusement park in Mexico City and, for two and a half years, at a much better facility at the Oregon Coast Aquarium. 'Lethargic' would be a fair, objective description of Keiko's behaviour when he arrived in Oregon. Can anyone truly believe that this wonderful, very family-oriented animal has led a life as full and rich as it would have been swimming with his mates in the vast depths and expanses of the oceans?

Jane Goodall has seen and tried to help all too many young chimps who have come to her Tchimpounga Chimpanzee Sanctuary after unknown traumatic experiences in childhood. One of these chimps hits himself repeatedly, all day long. Another, La Vieille ('The Old Woman'), has lived at Tchimpounga now for many years yet she is still afraid to leave her enclosure, so deep are the scars she brought with her. Anyone who looks into the eyes of this creature sees the buried sadness. Those who doubt are bluntly dismissed by the famous primatologist: 'I cannot imagine how anybody could possibly imagine that animals don't have personalities, feelings, emotions, and that their minds don't work very much like our own minds, especially chimps . . . I know they're capable of suffering.'

Goodall asks: why shouldn't we infer emotion when this inference seems warranted, just as I infer your anger or sadness when the situation makes this emotion not only plausible but likely? Goodall bases this judgment on almost 40 years' experience of living with chimpanzees. When one of the chimps at Tchimpounga jumps up and down and squeals in apparent pleasure, just the way a human child does, Goodall believes we should conclude that this pleasure is not just 'behaviour', but that it is felt as pleasure by the chimpanzee as well. When a young chimp reacts with alarm at a new pop-up toy, just as a human infant might, we should draw the same conclusion. Goodall knows whether a chimp's face is expressing pure relaxation, anger, sadness or embarrassment. She says bluntly, 'I defy anyone who knows anything about children to watch a small chimpanzee and not realise within a very short space of time that that child has exactly the same feelings, emotions, fears, despairs as human infants.'

Anything less, to Goodall's mind, is simply 'illogical', and certainly men and women who work professionally with animals have no choice but to see and carefully read the emotions of their animals. Correct judgment can be a matter of life and death with dangerous animals, and it often means the difference between success and failure with others.

Animals get bored and distracted. For professionals working with them, anthropomorphism (if that's what it is), regarding emotions, is sheer pragmatism. This does not mean, however, that these professionals understand the emotions of their animals as they would understand the emotions of their friends. With Irene Pepperberg's parrot Alex, the presence of powerful emotions is unmistakable, but often Pepperberg does not quite know what to make of them, or how to define them. Her model-rival protocol (which I explained earlier) is partly based on this species' renowned ability to become jealous, although, when she's being scientifically correct, Pepperberg doesn't use the term 'jealousy'. But she does know that if she cuddles another bird, Alex is likely to attack, or try to. When she sees in the morning a certain puffy-feathers, slit-eyes look, she knows she will get bitten if she makes a wrong move with her hand. But whether this is the bird's anger with her, or frustration over something else, she does not know. Then again, it's not all trial and tribulation with Alex; sometimes in the morning he has a certain relaxed posture, indicating that a productive lesson is in the offing.

She Knows Fear

Temple Grandin, Professor of Animal Sciences at Colorado State University, thinks in pictures, not in words, and she may therefore come much closer than the rest of us to understanding how animals really do see and feel about the world. In fact, she almost considers herself a member of a unique species, the autistic. In her essay 'Thinking Like Animals' in *Intimate Nature: The Bond Between Women and Animals*, she writes, 'Autistic emotion may be more like an animal's. Fear is the dominant emotion in both autistic people and prey animals such as deer, cattle, and horses. My emotions are simple and straightforward; like an animal's my emotions are not deep-seated. They may be intense while I am experiencing them but they will subside like an afternoon thunderstorm.'

Knowing that autistic children often seek the confines of a couch beneath the cushions, and that therapists use deep pressure to soothe autistic children, she inferred that cattle, specifically, would be soothed by soft pressure applied to wide areas of the body. She already knew they would be panicked by yelling and poking and prodding. She knew that curved chutes at the slaughterhouse would be more comfortable for the animals than straight ones, because the curve would hide from view the commotion at the end. She knew that blood horrifies people but is of no concern to cows. Instead, they worry about abrupt motion that may indicate a predator; they worry about 'high-pitched noises, disturbances of the dirt . . . a shiny reflection, or anything that appears to be out of place'.

Today, about half of all the cattle in North America spend their final hours in facilities designed by Grandin, author of *Emergence: Labeled Autistic* and *Thinking in Pictures*. Perhaps the reader is wondering what I wondered when I first encountered her ideas. Grandin replies:

I am often asked how I can care about animals and be involved in their slaughter. People forget that nature can be harsh. Death at the slaughter plant is quicker and less painful than death in the wild. Lions dining on the guts of a wild animal is much worse in my opinion. The animals we raise for food would never have lived at all if we had not raised them . . . It is important that our relationship with farm animals is reciprocal. We owe animals a decent life and a painless death.

It's Possible

While some people doubt the subjective nature of animals' emotions, others wonder whether their emotions are perhaps *more* powerful than our own, because they are less mediated by all the abstraction and rationalisation going on in the adult human's neocortex. Or, perhaps that is another way of saying that animals' emotions may be more childlike and unrestrained. When I see my three beautiful labradors racing around when we arrive at our cabin in the woods, I see pure joy. Here these wonderful dogs can run off the leash and swim in the stream to their hearts' content. I don't think I have experienced anything close to their unmitigated joy since I was a little boy arriving for the first time at one of my parents' rented beach houses.

No emotion for animals or more emotion? If that's the choice, I have to believe the latter.

Playing Around

In general, it seems to me, researchers tend to focus on the 'negative' emotions of animals, probably because more of us are inclined to agree that, yes, a lot of fairly basic animals experience fear and anger. We tend to be a little more possessive of the positive emotions of pleasure, happiness and joy, but maybe this isn't justified. After all, play is widespread in the animal kingdom, especially among mammals and birds. But is play really a manifestation of pleasure? That depends on how you define the word, and that turns out to be a tough one, of course. One author has compiled a list of 39 definitions! One of the more restrictive analyses I read suggests that play is simply the result of low levels of 'arousal' (not sexual arousal, but overall arousal). When bored, we play in some way. This does not sound very exciting, but such a definition actually bodes well for animals, because only a thinking being is capable of boredom. The oyster, we presume, does not get bored. A lot of other animals do.

Elizabeth Marshall Thomas, author of the bestseller *The Hidden Life of Dogs*, tells the story of a young dog of her acquaintance who had been given to an inactive, elderly couple who already owned an inactive, elderly dog who did not have the energy to keep up with the newcomer. The young dog had company, technically speaking, but no one to really play with. One evening Thomas saw this dog on a snowy hillside near his house, running in a circle with his nose to the ground, then stopping to burrow, then taking off again in the circle, then returning to the same spot for more burrowing. Six times the dog ran the route in this manner, 'alert, excited, tail high and waving, nose to earth'. In the beginning, Thomas figured some little rodent was the prey, but after the repeated loops she began to wonder and investigated the site, where she found no rodent hole, no rodent, nothing whatsoever. The dog had been playing make-believe, Thomas decided.

Any number of times I have watched one of my dogs stalk, presumably, some ferocious prey. A field mouse? A squirrel? Often this fearsome prey turns out to be a leaf. Somebody was playing make-believe, which, unfortunately, is not always the case, as my dogs' encounters with numerous skunks and two porcupines attest.

With dogs, the invitation for another dog or a human to play is the familiar 'play bow': forelegs flat on the ground, rear end high in the air, mouth wide open, tail waving like mad. This posture is unmistakable to us as well as to other dogs. In fact, the purpose of play may well be the development of social skills, but wouldn't it be nice if it had no developmental purpose whatsoever? After all, kittens play in the most charming ways, whereas cats are essentially solitary predators.

A more accurate interpretation might be that play behaviour is a rehearsal, under relatively safe conditions, for the real thing, life as an adult. (Is play anything else for children?) Play is, developmentally speaking, a serious business; play fighting is one of the most common forms of play. This plausible and attractive theory was first developed by Karl Groos in the late nineteenth century. By this definition, play exercises all the cognitive and motor functions the creature is, or will be, capable of: thus the kitten's instinctive stalking, sudden changes of direction, and bursts of speed. (These predator-like actions I recognised from my own observation of kittens. Not until I read *The Animal Mind* did I realise that I had also been watching rudimentary disembowelling actions.) Experiments with rhesus monkeys – experiments which cannot be condoned – have proved that animals raised with everything they need except playmates inevitably grow up to be dysfunctional adults – and they may go around doing a lot of disembowelling.

Generally speaking, the species we tend to think of as 'cleverer' are the ones that play – animals with 'a budget for exploration and self-stimulation', as philosopher Daniel Dennett quips in *Consciousness Explained*. These are mammals and birds. Off the coast of western Australia, dolphins surf the giant waves with apparent abandon. They are the only dolphins in the world known to do this. What possible reason could there be for this behaviour other than the sheer joy of it? And why haven't other dolphins 'discovered' this form of play? Spinner dolphins frolic in the wakes of large power boats, but they haven't been observed surfing natural waves.

One of Jane Goodall's first discoveries at the Tchimpounga Sanctuary was the chimpanzees' fondness for raiding her camp for new toys; another early discovery was the compatibility of puppies and chimps as playmates. I have seen a wonderful picture of a young bonobo playing blindman's bluff. There's no question that's what he's doing. Kanzi, the lexigram-trained bonobo in Atlanta, once hid from his increasingly worried trainer while flattened out on the bed covered with blankets – motionless – for 20 minutes! Then he burst forth in high spirits. Pure devilment – pure play, let's face it. And bonobos and chimps are famous for playing games of make-believe.

In *Elephant Memories*, Cynthia Moss's fascinating book about the 680 elephants living in the Amboseli National Park in Kenya, she writes about being 'trapped' in her Land Rover by a group of young elephants who 'rumbled, flapped their ears, shook their heads and made mock charges at me. I could only laugh at them, it was so obvious they were not serious.' A seriously angry elephant is unmistakable; these kids were playing.

Birds are harder to observe playing, but they do, and it's no surprise that some of the main avian players are parrots, ravens and crows. Parrots play with objects while lying on their backs. And, to all appearances, crows and ravens get a kick out of teasing gullible dogs. At least, they do so over and over and over again, with nothing to gain by it. Miriam Rothschild talks about her magpie who enjoyed mimicking ducks in order to get her dogs to dash out of the house in an attempt to catch them. The bird would preen mightily after this successful hoax. Paul Shepard tells of the hawks he observed while working as a seasonal park ranger and naturalist at Crater Lake National Park. These birds were circling around each other, the higher one dropping a wad of lichen to the lower one, who would catch it with his talons, circle above his mate and drop the lichen to him. Around and around they went. Shepard watched this game for several cycles before one of the hawks changed the game by flying over and dropping the lichen to him! (When Shepard told this story to a famous ornithologist who was also working in the park that summer, the man apparently snorted, turned abruptly and walked off. A behaviourist, no doubt.)

We have seen some of the amazing things honeybees can do, but can they also play? This is not a notion that comes to us naturally, but on sunny afternoons young worker bees a couple of weeks old venture forth for some simple loop-the-loops near the hive. Over the next few days they will venture farther from the hive, probably getting the hang of their navigational and mental-mapping skills, including calibration with the sun. Sounds like rehearsal play to me, but if this is indeed the point, it is not all that successful, as the Goulds point out, in *The Animal Mind*: honeybees suffer a 50 per cent attrition rate on their first official day of foraging.

Singing Gorillas and the Higher Emotions

In the wild, gorillas sing. They especially like to sing after a fierce thunderstorm has ended, the sky has cleared and the rainforest air smells sweet. Their singing is not of Pavarotti quality but it definitely sounds like singing and it certainly sounds as if it is coming from a happy animal. They do not sing in captivity, as a rule, but Michael, a resident at the California Gorilla Foundation, listens to Pavarotti for hours on end. He can hardly contain himself while tapping to the rhythm, utterly transfixed. Only the most jaded of observers would doubt that Pavarotti's singing makes Michael happy – just as it does millions of other fans.

Sue Savage-Rumbaugh tells the following story about her star pupil with the lexigrams, Kanzi. The bonobo's stepmother, Matata, had been away from Kanzi's quarters on breeding leave, which was successful. Now pregnant, she was due to return to the laboratory. Kanzi had done very little with the lexigrams before Matata's departure, but he had made great strides in her absence. What would happen now? Keeping the two separate 'for the sake of science' seemed difficult in terms of practicality and cruel besides, but Savage-Rumbaugh was worried that her star pupil would lose interest in his language studies now that his stepmother was back. She decided to reintroduce the two, then keep them separate for a few days and assess Kanzi's frame of mind.

On the afternoon Matata arrived, Kanzi returned home from a long, hot day in the forest. Savage-Rumbaugh immediately informed him that there was a 'surprise' waiting in the colony room. Kanzi expected food, naturally. Savage-Rumbaugh replied, 'No food sur-prise. Matata surprise. Matata in colony room.' Savage-Rumbaugh picks up the story:

> He looked stunned, stared at me intently, and then ran to the colony room door, gesturing urgently for me to open it. When mother and son saw each other, they emitted earsplitting shrieks of excitement and joy and rushed to the wire that separated them. They both pushed their hands through the wire. To touch the other as best they could . . . I hadn't the heart to keep them apart any longer, and opened the connecting door. Kanzi leaped into Matata's arms, and they screamed and hugged for fully five minutes, and then stepped back to gaze at each other in happiness . . . If we've lost him back to the world of the bonobos, so be it, I thought to myself.

But all's well that ends well. Kanzi gestured for his lexigram keyboard and indicated 'open'. If anything, Matata's return inspired him to greater glory with his lexigrams.

Grief in animals may be the hardest emotion to understand, at least in my experience. When my golden retriever Jenny died, her four-year-old daughter, Mink, looked and sniffed at Jenny's body and clearly did not have the slightest idea what to make of it. When we took Jenny's body outside to the grave we dug behind the house, Mink acted as though she had lost her mind. She ran and ran around and around the yard, over and over again until finally she dropped from sheer exhaustion. When we came back into the house, Mink would not leave us for a moment – pushing up against our legs no matter where we walked. Apparently, however, her grief, her realisation, did not hit her with its full force for two or three days, when she just stopped eating, lay down and paid no attention to anyone or anything. This went on for about two weeks, at which point we decided we had to get a puppy for Mink.

When the new puppy, Maggie, a tiny yellow labrador, arrived,

Mink showed only mild interest in her and we worried that this was not going to work – until the next morning. During the night these two bonded, became inseparable and remained that way until Mink died many years later.

Unlike dogs and many other animals, primates appear to show their grief immediately – especially if it is a mother who has lost a newborn. One of our most moving *Nature* films ever was about a group of Japanese macaques which had been filmed over several years by Japanese researchers. Our star was Mozu, and we called the film *Mozu the Snow Monkey*. During the filming, one of Mozu's relatives gave birth to a stillborn baby. The mother would not leave her dead baby – in fact she would not let go of it. She carried its little body everywhere.

At night she would climb up into a particular tree, hold the dead baby in her arms and scream. It was the most blood-curdling scream I have ever heard. She did this for three days and three nights. On the fourth day, she put the body of her baby on the jungle floor and disappeared into the jungle herself.

Joy is a higher-order emotion. So are compassion and empathy. To us laypeople, these terms 'compassion' and 'empathy' are closely related; in the minds of some ethologists, however, they are critically different. Compassion they think of as straightforward fellow-feeling, while empathy suggests a 'higher order' of identification with the feelings of another. But how could a creature show compassion without some sense of the feeling in the other creature that merits this compassion? That's too fine a line for me.

We have all heard stories about dolphins coming to the rescue of struggling human swimmers, stories we have every reason to believe because dolphins often help their own in similar straits, pushing an ailing fellow to the surface to breathe and holding the creature there for long periods of time. Clearly, dolphins must feel empathy for other dolphins and, apparently, for humans as well.

In *Bonobo: The Forgotten Ape*, Frans de Waal tells a series of remarkable stories about these primates that can only be interpreted as showing compassion and empathy. One features the famous Kanzi on videotape helping his minimally trained sister, Tamuli, deal with spoken English. When asked by Savage-Rumbaugh to groom her brother, Tamuli hesitates. Kanzi then takes her hand, puts it between his chin and chest, and stares at her with a questioning gaze. In another story, a mother sucks water into her mouth and delivers it to her whimpering two-year-old. She did this three times. In the Milwaukee Zoo an elderly male bonobo named Kidogo suffers from heart disease and has a great deal of trouble getting around. With one exception, the rest of the bonobos go out of their way to help this patriarch, and sometimes, when the one misanthropic five-year-old tries to tease him, another bonobo steps in to protect him.

Jane Goodall has seen many cases of adoption among wild chimpanzees, even adoption by adult males. Her favourite among these tales was the adoption of a sickly infant named Mel by rough, tough, 12-year-old Spindle, who let the youngster ride on his back, cling to his belly, and climb into his nest at night. Spindle even rescued Mel from hazardous encounters with sexually aroused males, encounters that earned Spindle rebuffs for his troubles. But he endured them for the sake of Mel. Goodall says there is no doubt that Spindle saved Mel's life by adopting him. Even La Vielle, emotionally scarred for life, apparently, cared for a young orphan even more needy than herself. And Goodall describes the mourning of the apparently healthy male chimpanzee Flint when his mother Flo died. He sat with the body for hours. He became more withdrawn by the day. And within a month he, too, died.

A Denver, Colorado, television station videotaped the next story and it soon became known around the world. A great African bull elephant was down on his side and either unwilling or unable to stand up. Elephants will always die soon if they cannot stand up. In this case, even a large crane was unable to pull the elephant up onto four legs. But, as it happened, a group of smaller, Asian elephants was led past the downed African elephant. These Asian elephants immediately approached their 'cousin' and prodded him into standing up, then supported him until he had the situation under his own control.

Empathy? Grief? These are emotions that other animals do not share with humans?

Of course, one could interpret all these situations and many others besides as something less than meets the eye. Behaviourists and quasi-behaviourists hold that all empathy is, at root, just as selfish as everything else we do. Even humans, much less other animals, are altruistic because we expect something else in return ('reciprocal altruism') or because it feels good to be altruistic.

Fine. If this is your argument, who can refute it? As Mr Spock might have said, it is not *susceptible* to refutation. But I would argue that *whatever* compassion and empathy are with humans, so they are with chimps and bonobos and dolphins and dogs – and a host of other animals that humans have loved. De Waal suggests that the bonobo might be the most empathetic of all the apes. Who's measuring?

Stories of compassion *across* species lines are not rare, either. I have already related the remarkable story told by Jeffrey Masson in *When Elephants Weep* about the dog and the lion, a friendship that was ultimately broken up, alas, by the suddenly mature lion.

In 1984, the famous gorilla Koko made headlines when her pet kitten, who was called 'All Ball', slipped out of a doorway and was run over by a car. Koko cried when Francis Patterson informed her of the death. Three days later, Koko signed 'cry' when asked if she wanted to talk about the kitten. Five years later, when shown a picture of herself and All Ball, she signed 'that bad frown sorry inattention'. (Before the unexpected death of All Ball, Koko had signed 'trouble, old' when asked why gorillas die, and she signed 'comfortable hole bye' when asked where gorillas go when they die.)

De Waal relates an amazing story about the seven-year-old bonobo Kuni and an injured starling in the Twycross Zoo in England. The keeper was worried that the bonobo meant the bird harm, but in fact Kuni went to extraordinary lengths to try to get the bird airborne once again. At one point she climbed to the highest point of the highest tree in the vicinity and, hanging onto the tree with both legs, unfolded the bird's wings, one with each hand, and threw it aloft. No luck, the starling plummeted to the ground, whereupon Kuni took up guard duty – apparently successfully, because by the end of the day the bird was gone.

Jeffrey Masson interprets the infamous Clever Hans episode as an instance of amazing empathy across the species – empathy by the horse, who could apparently read the most minute of inadvertent cues from its teacher. Hans would be the one who could teach us something, if only we had the senses to receive his messages. Masson argues that the mutual love of dogs and humans is *prima facie* evidence of the extraordinary empathy of each species for the other. Otherwise, how could there be such love? We don't feel such emotion for trees or for insects or for hamsters as we do for horses and dogs and cats. Maybe our intuition and our empathy 'know' something that we don't. It is a powerful argument.

Cats, although currently the most popular pet in America, *are* different from dogs, and cat lovers say, 'Amen!' Some pet lovers love both cats and dogs; some hate one and love the other. As with religion, it is best not to engage in a discussion about the relative merits of cats and dogs in polite society. In his book, *Dogs Never Lie About Love*, Masson relates the story of a German ethologist who argues that we chose dogs – wolves, at the time – for domestication, but cats chose us, while remaining true to their independent nature. The cat is *domestic*, according to this line of reasoning, but not truly *domesticated*. Those of us who like cats tend to think of them as little lions and tigers and cheetahs and love the wildness and independence in them. It's all part of their great attraction.

A Reciprocal Relationship

For centuries animals have been considered good medicine for ailing humans. Paul Shepard believes 'animal-facilitated therapy' began with the Quakers two centuries ago in England, where the inmates of an insane asylum lived with a garden, rabbits and poultry. In his excellent book *In the Company of Animals*, James Serpell reports the results of a study in Australia regarding the effect of the mere presence of a gentle dog in a nursing home for the elderly. For the 'experiment', two groups of residents were set up, one with whom the dog would live for six months, one which would carry on as it always had, without a dog. The difference in the mental health of the residents at the end of the test period was extraordinary. The residents who lived with the dog were healthier and happier. Serpell acknowledges that this test does not qualify as hard science. As with most of the studies in this area, this one could be criticised for its lack of strict protocols – the staff members assessing the success of the project were in favour of it in the first place. But there have been many other related studies, and I have not read of one that did not come to a similar conclusion.

A particularly surprising result obtained in Maryland showed that pet owners had a significantly better chance of surviving heart attacks. At some point, if there is any justice for animals, anecdote may become evidence. (I hope I live to see the day. But they had better hurry; I'm not a spring chicken any more!)

A series of tests conducted at the Johns Hopkins Medical School in the 1970s demonstrated the reciprocal benefits humans can provide for dogs. Serpell spares us the details, but the gist of the experiments was the measurement of heart rate and blood pressure in dogs subjected to electric shocks. Unsurprisingly, the presence of a petting human lowered the rate of increase in these two indices by 50 per cent. The presence of a human, combined with training to anticipate the shock, prevented the dogs from exhibiting any measurable stress, at least as measured by heart rate and blood pressure. (Serpell labels these experiments 'ethically disturbing'. I would call them unconscionable.)

The single most amazing such story I have ever heard about – a story that would make anyone pause and wonder – is from the anthology *Intimate Nature: The Bond Between Women and Animals*. The author is

Fran Peavey, 'writer, environmental activist, and peacemaker'. With permission I quote this brief chapter in its entirety, and without comment, because none seems necessary:

> One day I was walking through the Stanford University campus with a friend when I saw a crowd of people with cameras and video equipment on a little hillside. They were clustered around a pair of chimpanzees – a male running loose and a female on a chain about twenty-five feet long. It turned out that the male was from Marine World and the female was being studied for something or other at Stanford. The spectators were scientists and publicity people trying to get them to mate. The male was eager. He grunted and grabbed the female's chain and tugged. She whimpered and backed away. He pulled again. She pulled back. Watching the chimps' faces, I began to feel sympathy for the female.
>
> Suddenly the female chimp yanked her chain out of the male's grasp. To my amazement, she walked through the crowd, straight over to me, and took my hand. Then she led me across the circle to the only other two women in the crowd, and she joined hands with one of them. The three of us stood together in a circle. I remember the feeling of that rough palm against mine. The little chimp had recognised us and reached out across all the years of evolution to form her own support group.

Elephants

At birth, the brain of almost every mammal is already about 90 per cent of its adult weight. Humans are exceptional, with a brain at birth only about 25 per cent of adult weight. Elephants are right behind us, with brains about 35 per cent of adult weight. As Cynthia Moss writes, they have 'a lot of mental and physical development ahead of

[them]'. So it stands to reason – and there is plenty of evidence as well – that elephants may have as rich an emotional life as any creature on earth. Moss, however, is not interested in comparing the family life of elephants with our own. She looks at elephant life for its own sake.

Moss, whom I consider a kindred spirit and friend and for whom I have the greatest admiration, describes a chaotic meeting between two large groups of elephants, complete with 'earflapping, rumbling, screaming, trumpeting, clicking of tusks together, entwining of trunks, spinning and backing, and urinating and defecating'. This went on for four or five minutes. Describing another greeting later between two other groups, Moss writes, in *Elephant Memories*, 'Above all, the sounds of their greeting rent the air as over and over they gave forth rich grumbles and piercing trumpets of joy.' The author concludes:

> After 18 years of watching elephants I still feel a tremendous thrill at witnessing a greeting ceremony. Somehow it epito-mizes what makes elephants so special and interesting. I have no doubt even in my most scientifically rigorous moments that the elephants are experiencing joy when they find each other again. It may not be similar to human joy or even comparable, but it is elephantine joy and it plays a very important part in their whole social system.

Like Cynthia Moss, Joyce Poole has studied elephants for most of her professional life. In fact, Poole was a student at Moss's camp in Kenya and has written her own book on the elephants of Amboseli – *Coming of Age with Elephants* – and she, too, is in awe of the greeting ceremonies. She has even concluded that the family life of an elephant may be richer emotionally than our own and she sees this as a result of the incredibly close matriarchal society of female elephants: 'Imagine being born into and living with a family until the day you died, living with your sisters, your aunts, your grandmother, your mother, never straying more than ten meters [away]. When anything happened that threatened that family, or anything joyful happened and you came together as a unit and screamed and yelled and sort of threw your arms around each other — that's what it's like living in an elephant family.' With elephants, the maternal bond gives every indication of being just as strong as with us, and lasts a lifetime as well, but only with the females.

Males eventually go off on their own and may never return. For about eight months a year, they hole up in a bullpen, almost literally 'hanging out with all the other males', as Moss explains, getting fit, building strength for musth season. Since a mature female may bear a child only every fourth year or so, and there are probably only a few mature females in any one family, the male would be 'wasting a lot of his time', reproductively speaking, by sticking with just one family. He is better off trying to become the big alpha male for a large territory, in which case he could father as many as 100 calves in those four years — evolutionary permission to be as promiscuous as possible.

In her book *Elephant Memories,* Moss describes the heart-rending sight of a family of female elephants trying to save Tina, who had been shot in the lungs by poachers. When the confrontation with the men ended, the elephants turned to the downed animal. (One elephant had already died with a bullet in the brain.) 'Her knees started to buckle and she began to go down,' Moss writes, 'but Teresia got on one side of her and Trista on the other and they both leaned in and held her up'. Teresia was Tina's mother. 'Soon, however, she had no strength and she slipped beneath them and fell onto her side. More blood gushed from her mouth and with a shudder she died. Teresia and Trista became frantic and knelt down and tried to lift her up. They worked their tusks under her back and under her head. At one point they succeeded in lifting her into a sitting position but her body flopped back down.' Other elephants tried to stuff grass into Tina's mouth. 'Finally . . . when [Teresia] got to a standing position with the full weight of Tina's head and front quarters on her tusks, there was a sharp cracking sound and Teresia dropped the carcass as her right tusk fell to the ground. She had broken it a few inches from the lip well into the nerve cavity, and a jagged bit of ivory and the bloody pulp were all that remained.'

They gave up then but did not leave. Tina's family buried her with leaves and earth in the shallowest of graves. They stood vigil through the night, then began to move off. 'Teresia was the last to leave,' Moss writes. 'She reached behind her and gently felt the carcass with her hind foot repeatedly. The others rumbled again and very slowly, touching the tip of her trunk to her broken tusk, Teresia moved off to join them.'

On one other occasion Moss witnessed what appeared to be the beginning of a burial ceremony – the elephants had broken off branches and palm fronds and placed them on the carcass – before a spotter plane flown by park rangers swooped down and inadvertently interrupted them. Moss learned that the myth of an elephant graveyard is just that, a myth, although there are certain locations with good food and water that attract ailing elephants and might therefore have more than their share of bones. She suspects the myth arose following the discovery of enormous piles of bones remaining from the slaughter of an extended family of elephants. In southern Africa, one tribe killed entire herds by building a ring of fire around them. Although the graveyard is a myth, Moss believes the animals do have a concept of death. She saw too many episodes like that between Teresia and her daughter.

She writes:

> Although they pay no attention to the remains of other species, they always react to the body of a dead elephant. I have been with elephant families many times when this has happened . . . they stop and become quiet and yet tense in a different way from anything I have seen in other situations. First they reach their trunks toward the body to smell it, and then they approach slowly and cautiously and begin to touch the bones, sometimes lifting them and turning them with their feet and trunks. They seem particularly interested in the head and tusks. They run their trunk tips along the tusks and lower jaw and feel in all the crevices and hollows in the skull. I would guess they are trying to recognize the individual.

Joyce Poole adds that elephants often greet one another by putting their trunks in each other's mouths. She believes this behaviour is related to the behaviour with bones, when, as Poole says, 'They're very, very quiet. Terribly quiet. In fact the only thing you hear is the clank of bones and the sounds of the elephants exhaling, and the reason they're exhaling is so they can take in another deep breath so they can smell the bones.' Poole believes it is very possible that elephants do recognise the bones of family.

Anyone who has worked with captive elephants has stories confirming the old cliché that an elephant never forgets. Many involve a particular keeper who has dropped out of involvement with an elephant for a decade or more, maybe even for two decades. The moment this keeper returns to the scene, the elephant reacts favourably or unfavourably to the memory, as the case may be.

The elephant behaviour with bones is so predictable that wildlife photographers and film-makers can easily 'bait' a scene by bringing in bones from another site. Perhaps none of this should surprise us, because the close-knit, matriarchal society of elephants can be devastated by the loss of the matriarch and never recover its cohesiveness. Moss has seen a mother wander lethargically at the rear of the family for many days following the death of her calf.

How ironic it is that Moss has so often witnessed elephants going out of their way not to kill a human, despite every kind of thoughtless, mean provocation. 'At times,' she writes, in *Elephant Memories*, 'it was actually difficult for an elephant not to step on or run over someone but they always swerved or backed quickly to avoid doing so'. She adds that this behaviour is not necessarily out of consideration for us. Elephants are not predators, after all. They are herbivores, not carnivores. They have nothing to gain in the wild by killing an animal. On the other hand, elephants can definitely be dangerous in captivity, because, Moss explained, the human–elephant relationship in captivity is one of dominance, with the human as the dominant creature. If the elephant sees the opportunity to even the score, it will. In fact, the elephant, male or female, actively seeks any opportunity to redress old grievances – which the elephant does not forget. This is why several keepers around the world are killed every year — one of the highest per capita fatality rates for any occupation.

An elephant does not need to tusk or stomp in order to kill. From time to time they have killed cows weighing 450 kg (1000 lb) with one swipe of the trunk. They well know their immense strength. The danger involved in working with elephants, and the direct correlation between this danger and the dominance relationship, has led to the theory that dominance should be avoided in the first place. The elephants should be treated as equals at all times. Common sense, we might think, but many keepers are reluctant to give up the old way, even though some day they may pay for it with their lives.

And equality is all they ask for. Just think what damage they could do, as the largest land animals on earth, if evolution had moulded them to think of themselves as superior to all other living creatures.

chapter 11

CONSCIOUSNESS

The Self-conscious Emotions

The primary emotions considered in the previous chapter — fear, anger, pleasures of all sorts — are simple, powerful, and shared, I believe, by many animals. This is the parsimonious, the least embellished, conclusion (although behaviourists would not agree). However, the secondary emotions — of embarrassment, guilt and shame — are considered by psychologists to indicate a higher order of consciousness. Apparently, these emotions require, on the part of the person or animal, a sense of 'what I look like' to someone else. This in turn seems to require a sense of myself and a sense that others might have an opinion about me. In short, only creatures with both 'other-consciousness' and 'self-consciousness' would seem to be 'capable' of embarrassment and shame. Otherwise, why and how would a creature care about another's opinion? These are therefore called the 'social' or 'self-conscious' emotions.

Most animals give no indication of experiencing shame or embarrassment. No matter what a creature might be capable of, in terms of thinking and in experiencing fear and pleasure, we do not really expect most animals to be self-consciously embarrassed about anything. At least I don't look for it from a diamondback rattlesnake or from a largemouth bass. No matter how complex the social organisation within a beehive, we do not imagine the forager bee who has returned empty-handed, so to speak, being ashamed of herself — but it is unlikely that we would be able to tell for sure.

We do not have easy answers concerning birds, either. Irene Pepperberg is pretty sure that when Alex the parrot says 'I'm sorry' (which he does quite often, and with good reason) that he is not really sorry at all. He will use the phrase to defuse what he correctly reads is her annoyance at his behaviour. Pepperberg, however, does not sense any genuine embarrassment and contrition, and the bird will repeat the annoying behaviour almost instantaneously. These birds are extremely self-centred. All in all, Pepperberg puts Alex at the emotional level of a two-and-a-half-year-old child — maybe a three-year-old, on his good days. She laughs when reporting that the standard books on raising parrots warn that to have one is to have a young child on your hands for the rest of the bird's life, which can be 40 or 50 or even more long years.

Neuroscience tells us that embarrassment and shame are the emotions of the neocortex. Every mammal has a neocortex sitting atop its brain, but the structure is not necessarily large, and most mammals show no evidence of the self-conscious emotions. When the nineteenth-century horse Clever Hans made a mistake, was he embarrassed in front of his audience? I doubt it. Most of us draw the line for the self-conscious emotions quite high.

Dogs? According to Jeffrey Masson, trainers of guide dogs say that old dogs who become incontinent do seem to be embarrassed. Most dog owners have seen 'submissive body language' in pet dogs after some 'elimination mistake' that certainly appears to be embarrassment and shame. My own dogs seem to be embarrassed after refusing to come when called or when scolded for 'stealing' food, as well as with the other problem.

Not everyone agrees with me on this, but I think cats, too, become embarrassed from time to time. When they are scolded for whatever reason, usually completely deserved, they slink off to a hiding place and often will not be heard from for hours. Perhaps the emotion they are feeling is anger, not shame. Who knows? Certainly cats *do* appear to be capable of pride when dropping dead birds and mice at your feet!

Embarrassment, guilt and shame serve as regulators of social life. That is the standard assessment, at any rate, as demonstrated in the work of Dr Michael Lewis, of the Institute of Child Development in New Jersey, and others. Children first show signs of embarrassment about the age of two. By this age infants have a sense of themselves, of others and of rules and expectations. In the *Nature* animal minds series, we see this demonstrated in a test rigged for a little boy to fail. When he does he responds with the 'collapsing' body language psychologists associate with shame: head and neck slumped, face turned away to hide from the eyewitness. (Pride produces exactly the opposite body language: shoulders up, chest out, head high, arms flung overhead.) Then we see Indah, an orangutan living and studying at the Think Tank at the National Zoo in Washington D.C., learning language with a lexigram keyboard. If the keyboard is rigged for Indah to fail, she exhibits the rudiments of the classic collapsing body we see with the small boy: throwing her arm over her head and ducking to the side.

'Kanzi be bad now.'

Thus did our favourite bonobo one day signal his caretaker Sue Savage-Rumbaugh before promptly swiping his half-sister Panbanisha's banana. But does Kanzi really feel bad? Savage-Rumbaugh was curious and set up a simple trial, the equivalent of a test which has been tried on children for many years. Using the lexigram keyboard, she informs Kanzi that certain new toys are very fragile, and that he must be careful when playing with them. Unknown to Kanzi, of course, these toys have been rigged to break. They do so when Savage-Rumbaugh is watching through a one-way mirror. When the toy breaks, Kanzi reacts as every human child reacts. First he looks suitably embarrassed. Then he glances around to see if anyone witnessed the transgression. Then he tries to put the toy back together. Unsuccessful, he finally hides it under a rug.

In humans, the artistic and religious impulses, which we believe are at least as old as the earliest cave drawings, are in some ways self-conscious emotions. Certainly they require self-consciousness of a high order. Jane Goodall reports that chimpanzees sometimes make threatening gestures against thunderstorms, a report interpreted elsewhere as a 'proto-religious' behaviour. She has witnessed chimps reacting to waterfalls in ways that, again, are impossible not to anthropomorphise about. Goodall says, 'I ask myself [whether] perhaps it was feelings of awe like this which led to the early primitive, animistic religions.'

One day at Gombe one of Goodall's assistants was seated on a ridge, watching the sun set over Lake Tanganyika. We can readily imagine how beautiful this scene was. Soon two chimpanzees climbed the ridge, where they greeted one another and the student, and then sat down in silence to contemplate the blazing colours of the sunset. There have been other reports of chimps watching sunsets. Perhaps some readers snigger at these stories, but I fail to see why. Many of the primates are, like humans, trichromats: we see three main colours – red, green and blue. If you and I see the sunset in the same way (and of course there is no proof that we do, if you want to be pedantic about it), then there is every reason to believe that a chimpanzee is also seeing the same sunset in the same glowing colours that we are. Many other mammals, including cats and dogs, are dichromats, 'missing' one of the three primaries, but there is no reason why a sunset shouldn't be a pleasant experience for them as well. Most of us have seen drawings by primates, abstractions pleasing to the eye and obviously pleasing to the artists themselves.

One of my most prized possessions is a gentle, pastel abstract painting by 'Mary the Painting Elephant', an abandoned circus Asian elephant who lives at Riddles Elephant Sanctuary in Arkansas. During a live broadcast on Arkansas Public Television, I stood next to Mary and her easel as she composed this work, using her trunk to hold the brushes, while she splattered watercolours all over my new suit as well as her canvas. No matter. I fell in love with Mary, and if the way she tousled my hair with her 'painty' trunk was any indication, I think she liked me a little bit, too.

Degrees of Consciousness? Why Not?

The 'higher' emotions allow us to ease into the subject of consciousness by way of the back door, because the front door is so impregnable! After all, consciousness is what it's all about! Thomas Nagel's famous question, 'What is it like to be a bat?', implies that it's like anything at all. If it's not, then the bat is, just as Descartes said, a machine made of nerves and muscles. Does it feel like something to the bat to be the bat? To the elephant to be the elephant? To the bee to be the bee? The line between those animals for whom it is like something, and those for whom it is not, is consciousness. Everything discussed so far has, in effect, been a prologue to this question, and I am convinced that we have good reasons for inferring from our own experience that consciousness in animals must be *something* like our own. I feel certain that it feels like something to be the elephant, certain that it feels like something very different to be the bat, and I believe it might well feel like something to be the bee — but, I cannot *prove* it.

Many of us intuitively have doubts about bees because of the instinct factor, but nowhere is it written that an animal, human or otherwise, cannot be conscious of an instinct. Sneezing is one of Donald Griffin's favourite examples. Sneezing is entirely reflexive on our part, yet we experience it quite clearly! Breathing is an instinct. Language is an instinct, in effect, but we experience language quite consciously. One could continue this list indefinitely. Therefore an animal's life does not have to be primarily a matter of versatile or even voluntary behaviour in order to be a conscious one.

If indeed consciousness arises from the activity of the myriad neurons in the brain, what difference does it make if this activity is instinctive or learned, simple or complex? Perhaps none at all — an argument that garners support from an unexpected quarter. As we have seen, Celia Heyes is an agnostic, at best, on the question of animal minds. In her daily life she, like all of us, is guilty of folk psychology and acts on presumptions of consciousness, but as a professional she thinks it is quite possible that no animals other than human beings possess consciousness. She definitely believes there is no way of ascertaining whether they do or not, and she discounts all inference arguments as being loaded with pitfalls. And yet this consciousness-doubter agrees when Donald Griffin argues against the intuitive connection we draw between complexity and consciousness. Heyes explains, 'Sometimes I may be gazing at a blank wall and I'm conscious of seeing the wall and of seeing the colour of the wall . . . The thinking is very simple, but I am just as conscious of it as I am when I'm thinking complex things.'

Nowhere is it written in stone that consciousness is an all-or-nothing deal. Maybe it's like a dimmer switch, to employ Oxford psychologist Susan Greenfield's useful analogy: held at a dim setting by some species, a little more brightly by others, and so on up the scale. In fact, degrees of consciousness are what we should perhaps expect. This is a vitally important point. Those who doubt that animals possess consciousness and emotions must explain how it is that these attributes arrive in the natural history of *Homo sapiens* like a bolt of lightning, with no precursors in the animal world. Such 'overnight success' simply does not make much sense, intuitively or biologically.

It is certainly true that there is an enormous difference between the human mind and any animal mind, between our language and any animal communication or language. Even Darwin, a staunch proponent of continuity of the species, was puzzled by how great the difference is between us and all the others. But our position on the pinnacle does not negate the need for precursors. As Allen and Bekoff write, in *Species of Mind*, 'We believe that arguments about evolutionary continuity are as applicable to the study of animal minds and brains as they are to comparative studies of kidneys, stomachs, and hearts.'

Ironically, perhaps, partial and damaged consciousness offer an excellent means of getting a better grip on the subject. The phenomenon of blindsight is a case in point. In Chapter 3, we looked at this rare phenomenon as an example of what cognition without consciousness might be like. It can also be used to illustrate partial consciousness. Briefly put, blindsight victims have a large blindspot in the field of vision, as a result of some sort of injury to the part of the brain known as the visual cortex. When both the left and right visual cortices have been damaged, the individual is completely blind. Yet some of these individuals can nevertheless perform quite well on tests involving the identification of simple shapes and flashes of light. They see, apparently, and they report on what they see, but they have no consciousness of doing so.

One such individual is Graham Y, who is one of Dr Larry Wisecrantz's patients at Oxford University. After a boyhood injury, Graham lost sight in his right field of vision. He made adjustments and went on with his life; unable to obtain a driver's licence, he cycled everywhere. Then it was discovered that this young man has blindsight. He 'sees' certain rapid movements and flashes of light, but others he doesn't see – or so he's told, because he has no sense of seeing any of them. 'It is very bizarre,' Graham says. On tests set up to assess his colour vision, he once again scores quite well, especially with the colour red, although he has no sense at all of seeing red. 'I find it just as bizarre as you do, to be quite honest,' he repeats. What does it feel like? 'Nothing.'

Animals can also have blindsight, which is not surprising, since the brains of mammals, particularly, have essentially the same structure as ours. Dr Nicholas Humphrey spent years working with the monkey Helen, whose visual cortices were not functioning. Yet Helen learned to see, in a way. But then if a stranger came into the room, she became blind once again, or at least stumbled into things as if she were blind. Humphrey thinks of blindsight as something akin to extra-sensory perception. He says, 'They have the information, they don't know where it came from and they find that very peculiar indeed.'

Other researchers conducted an important series of tests on a monkey who was, like Graham, blind in his right field of vision. This monkey was presented with two sets of green lights, one in the left field of vision, one in the right. He was trained to touch any light that was illuminated, and to touch a separate blue light if none of the green lights were lit. The results of the test confirm our expectations regarding blindsight. When a green light on the left (his 'good' side) was lit, the monkey touched it. When a green light on the right was lit, the monkey touched it *but also touched the blue button*. He saw the green light on the right, but he was not aware of seeing it, and therefore touched the blue button as well. He was conscious of the light in the left field of vision, but not the one in the right field. Some comparative psychologists believe that this test comes as close as any ever conducted to demonstrating animal consciousness in a controlled laboratory setting.

In the previous chapter we looked at Temple Grandin's sympathy with the emotions of animals. Grandin is autistic. Whatever the exact nature of this illness, autism definitely produces an inability to perceive and deal with the world at large in the 'normal' way. Autistic individuals do not synthesise their perceptions into an organised and coordinated picture of the world. As Grandin says, she has all the emotions, but they come and go quickly. As someone else put it, the autistic's life is not a rich, Tolstoyan novel, but something more like a minimalist short story. The autistic person has a different kind of consciousness from the rest of us. Although Grandin does use words, and very well, she thinks in pictures, not words. A cow also thinks in pictures, not words, and you will never convince Grandin that both she and the cow are unconscious.

Consciousness as Adaptation

Donald Griffin writes, 'If we postulate the presence of simple thoughts, a great deal of animal behavior can be understood as a consistent adaptive pattern.' He told me in his Harvard office, 'Animals seem to be conscious of things that are important to them. "That is something good to eat", or "That is something dangerous to get away from", or "That's a female of my species that's attractive to me", or "That's a baby that needs help or needs food." Those are the sorts of consciousness that seem likely.'

The philosopher Daniel Dennett suggests that this capacity for generalisation might evolve when the list of behaviours required of a species becomes unwieldy for 'if-then learning'. As he summarises the situation, all individuals in all species 'learn through evolution' – through inherited hardwiring. Then a surprising number of species, including simple invertebrates, find themselves compelled to learn in a second way (what Dennett refers to as 'Skinnerian fashion') – conditioned learning à la Skinner's behaviourism. And some select species are then pressured by evolution to learn in yet a third way, by using information in an inner environment (the 'mind') in order to draw conclusions about the outer environment. Here we have consciousness, pure and simple, and, as the philosopher Karl Popper puts it (quoted by Dennett), consciousness 'permits our hypotheses to die in our stead'.

This is a witty and elegant phrase, well worth a moment's consideration. I think of a hungry lioness hunkered behind a mound, considering a grazing buffalo 30 metres away. A buffalo is no wildebeest. A buffalo is a different matter altogether, one of the most aggressive and dangerous animals in Africa, and the lioness understands this. Yes, she is hungry and so are her cubs, but no, she is not that hungry and not that good – or stupid – a mother. She will let her hypothesis die in her stead. She strolls away. Or, think of all the fantasising we humans do in moments of fear or anger. Pretty horrible thoughts but we do not act on them.

Why can the great apes solve complex problems of analogical reasoning and achieve what they do with symbolic chips and lexigrams – talents they would appear not to need in the wild? The answer is one of the keys to this whole inquiry, really. The apes are clever enough to solve our abstract problems because they do in fact need to solve equivalent practical problems in the course of their own lives in the wild, where programmed instinct and 'Skinnerian' conditioning would not be nearly enough to see them through the day successfully. Consciousness quite possibly arose when social animals (a category that could include certain insects, don't forget) reached the point at which they were required to understand the intentions of the other animals.

In Darwinian terms, consciousness would thus be one final, vitally important, and beneficial adaptation. And thus would animals be 'natural psychologists', to employ the term first broached by Nicholas Humphrey, while elaborating on the hypothesis of Alison Jolly.

The same point was made more colourfully by the primatologists David Premack and Guy Woodruff, who wrote that they were less interested in the ape as a physicist than as a psychologist. And two more primatologists, Dorothy Cheney and Robert Seyfarth, write, in *How Monkeys See the World*, '. . . group life has exerted strong selective pressure on the ability of primates to form complex associations, to make transitive inferences, and even to judge causal relations, but primarily when the stimuli are other primates. Monkeys and apes are more likely to solve at least some types of problems if they involve social stimuli [from members of their own species] than if they involve objects.'

We saw in Chapter 10 how the intensely close, matriarchal society of the elephant clearly facilitates deep emotional reactions and bonds. According to Jane Goodall, the patriarchal society among chimpanzees applies pressure to develop a different set of social skills, including the recognition of individuals and the accurate assessment of relationships that are always in flux, with chimps moving in and out of the group on a daily basis. We would be extremely naive or wilfully obtuse not to acknowledge that handling such a complex society requires some commensurately complex social skills. Goodall says, 'They watch each other very carefully, they pick up tiny little intention movements, they intently look at one another's faces and seem able to read the expression and it's just very clear from the way the chimpanzee reacts, what he does or doesn't do, that he understands perfectly well the intention of the other.'

In his book *Chimpanzee Politics*, Frans de Waal tells the story of the two mothers, Jimmie and Tepel, watching their children playing. When the playing gets rough and mild admonitions do not quell the outbreak, Tepel gently prods a resting female, Mama, and points in the direction of the quarrelling kids. Mama is a high-ranking matriarch unrelated to the others. De Waal writes, 'As soon as Mama takes one threatening step forward, waves her arms in the air and barks loudly the children stop quarreling. Mama then lies down again and continues her siesta.'

De Waal believes that Jimmie and Tepel were concerned about starting conflict between themselves as a result of conflict between their kids, so they called in the third party, Mama, to restore order. This is an illustration of what has become known as a 'social tool'. Another example of the adaptive use of a social tool is the well-known behaviour in which one female primate will 'present' her rear end to a ranking male while threatening another female at the same time. Her idea is to enlist the male's help, perhaps to the point of inducing him to attack the other female. All too predictably, perhaps, the male is often taken in by this ploy.

There is literally no end to the stories of social intelligence involving primates. It is their – our – defining trait. One day in Georgia, Sue Savage-Rumbaugh was hiking through the woods with the bonobo Panbanisha and some dogs. (Panbanisha is the half-sister of Kanzi, whom we already know well.) Suddenly the dogs stopped in their tracks and began growling at some threat in the trees. Panbanisha and Savage-Rumbaugh soon spotted the source of the dogs' concern: a large wildcat perched on a limb. Both the bonobo's and the humans' hair instantly stood on end, and the two exchanged a telling glance of apprehension before immediately heading to safety. Back at the lab, Kanzi, his stepmother Matata, and another ape, Panzee, all understood immediately that something frightening had happened outside. They stared out of the window, making the soft 'whu-uh' sound that registers unusual events. Panbanisha vocalised, and they responded in kind. Savage-Rumbaugh spoke to Kanzi and Panzee in English, and 'both of them listened with rapt attention and huge round eyes. At appropriate points during my recounting Panbanisha embellished my tale with bonobo "Waa" vocalisations, as though to add her own emphasis.'

In the days that followed, Kanzi and Panzee were obviously concerned whenever they approached the area in which the big cat had been seen. By what means and to what extent we may never know, but they had understood the gist of the scary situation. Of this there is no doubt, not to Savage-Rumbaugh. She acknowledges that one such episode doesn't prove much, but when such episodes happen all the time – which they do, in her experience – she is left with no alternative but to conclude that highly effective and complex social interaction and communication are taking place.

Elizabeth Marshall Thomas, author of *The Hidden Life of Dogs*, tells us in that book about a visit to Baffin Island with a contingent of Canadian scientists. Their appearance disturbed the local pack of wolves, and the young wolf and his mother who first saw the human beings retreated to their den and howled. An hour later all the wolves had congregated and were howling in a haunting chorus. For wolves, Thomas speculates, howling is equivalent to our choral singing. It's a means of bonding and of communication – 'perhaps to express their solidarity vis-à-vis our group, or perhaps to tell us that the river system was owned by them and that we should move'. One of the scientists predicted that the other wolves were also being warned away from the site of the trespassing. And this was true. Thomas writes:

> Uncanny as it seems, as long as we stayed there, the wolves never again visited the place where they first spotted us, the place that became our main camp, even though we certainly didn't threaten them but instead put out very odoriferous baits to lure them. They later tolerated and even visited me in my camp right by their den, and a wolverine came gladly enough to eat the baits, but . . . the wolves thereafter avoided our main camp, just as if the two who had seen us told the others where a surprising and untoward thing had occurred. In fact, they almost certainly had done just that. None of us knew how they managed the communication, but then, wolf communication is merely one of many things about animals that people don't fully understand.

Although we don't tend to think about them in these terms, birds are also highly social creatures, and they are at their most intelligent in socialised settings. Irene Pepperberg emphasises this point regarding Alex and other African grey parrots, who in the wild live in large flocks for many years. Each of these birds has to process a lot of information about the environment and other birds. I've mentioned Pepperberg's model/rival, or M/R, learning environment, in which Alex's manifest capacity for jealousy is exploited for learning purposes. This M/R environment also exploits social interaction between the trainer and Alex and other parrots. Pepperberg knew what every teacher of young children knows, that actually handling an object helps the child to learn the name for this object. Labelling at a distance simply does not work well with children, and Pepperberg believed that the same would be true of parrots, and she was right. Like the child, the bird needs hands-on interaction.

Other Minds

This brief discussion of highly evolved social intelligence and 'social tools' — a discussion that could have been a book in itself — leads directly to the underlying question: do some animals have a 'theory of mind'? This is the official terminology, although the meaning of this phrase is more accurately conveyed as 'theory of other minds'. When Tepel pokes Mama, for example, does Tepel actually have a sense that Mama is a being with independent intentions and beliefs, with a 'mind of her own'? Or has Tepel simply learned from past episodes that poking Mama in this way results in a certain behaviour? When two lionesses intentionally make themselves known to a wildebeest, while a third and fourth sneak into position for a blindside attack, does each have a sense of the intention of the others?

The question is important because an animal's knowledge of other minds implies a knowledge of its own mind. Or does it? The authors of *How Monkeys See the World* believe that vervet monkeys master some of the abstract relationship tests because they encounter 'dominance hierarchies' in their real world every bit as complex as those of the Montagues and Capulets, but Cheney and Seyfarth also note, 'The fact that we can describe vervet monkey social behaviour in terms of relationships and strategies does not, therefore, necessarily mean that knowledge of these principles is what guides the monkeys' actions. They might be following a few relatively simple rules of action: avoid some individuals, groom or copulate with others. At this point in our discussion we simply do not know.'

Maybe the monkeys know only about other *behaviour*, not about other *minds*. Or, as philosopher Daniel Dennett puts the point, it is theoretically possible to be an unthinking natural psychologist. Certainly a theory of other minds would come in handy for any primate, so it would have good reason to evolve, but this is not to say that it necessarily has. An animal does not necessarily need to consult any internal 'model' of the mind in order to anticipate behaviour, Dennett points out. The animal could just be equipped with a large list of if-then contingencies.

On the other hand, as I've noted, at some point a lengthy list would need to be replaced by some simple generalisations and ideas. But enough of these theoretical debates! Let's look at some hard-and-fast research, beginning with the work of Premack and Woodruff. They asked if an intelligent animal such as a chimpanzee who knows what it's like to wear a blindfold understands what it's like for another chimp or a human trainer to wear a blindfold? To find out, they set up a situation in which a young chimpanzee required the assistance of a trainer in obtaining food. The chimp knew where the food was; the trainer had the key to the box. Four chimps participated in this study, and each had no trouble leading the trainer to the hidden food. Then one day the trainer arrived blindfolded and was unable to follow. In fact, the trainer had several 'blindfolds': one across his eyes, others elsewhere across his face.

What would each of the chimps do now? The results were fascinating: three of the four chimps took the trainer by the hand and physically led him to the box. Very logical, very productive: good common sense. The fourth chimp, however, took off the blindfold —

the correct blindfold — leaving the other decoys in place. Does this prove that the fourth chimp understood about his own eyesight and about the trainer's eyesight — knew about his own consciousness and the trainer's as well? Perhaps, but one could also argue that the chimp simply didn't like the appearance of the blindfold; one could argue that the chimp wanted to be able to see the trainer's eyes, and by removing the blindfold thereby stumbled into performing an intelligent-looking action. In this field, one can always argue.

In another Premack and Woodruff study, the chimpanzees dealt with two trainers, one of whom was established over time as 'friendly' (he shared food) and the other as 'unfriendly' (he did not share the food, and he also dressed like Darth Vader, which the chimps did not like). In time, the chimps learned to direct the friendly trainer to a container of food, and one of the chimps learned to deceive the unfriendly trainer. But, again, the question arises: did the chimp truly have a sense of the mental state of the friendly versus unfriendly trainer, or did he just learn the simple rule 'unfriendly trainer equals no food'?

To my mind, a much better case for a concept of other minds is put forward by Sue Savage-Rumbaugh. In her laboratory in Atlanta, the chimpanzees Austin and Sherman learned to work with each other and their lexigrams to pass food and tools back and forth in carefully controlled situations. One of the two watched as a trainer put food into one of several containers, while the one who could not see the container controlled the tools needed to open them. What to do now?! After several weeks, the chimp caught on and began asking the other, by way of the lexigram keyboard, to pass the tool, which was done. I discussed this work in the chapter on language, but it is even more significant in relation to the question of other minds. As Savage-Rumbaugh writes, in *Kanzi: The Ape at the Brink of the Human Mind*:

> Did Sherman know that Austin's mind or knowledge state was different from his own? It seemed so; otherwise, why would he bother to tell Austin which tool he needed or which food was hidden in a container? . . . It was the chimps who began

to attend to each other, to coordinate their communications, and to correct each other's errors. For me, as I watched them, the question was reversed: How could they not know that they were communicating?

Another study I find compelling involves a chimp observing while a trainer loads one of four containers with a favourite food. That is, the chimp observes the baiting procedure but cannot see which of the four containers has the food. But she can see two trainers who are also observing the baiting. One of these trainers (the 'knower') has a clear view, while the view of the other (the 'guesser') is blocked by a screen. The challenge is to seek the help of one of the trainers in finding the food. Of four chimps tested, two chose the knower, indicating an understanding of the 'seeing–knowing' relationship for another individual. One chimp chose this knowing trainer but then did not always follow her advice. The fourth chimp never caught on to the difference between knowing and guessing.

Other versions of this and similar tests have produced the same results: some apes catch on, some don't. Similar tests with macaque monkeys failed to instil knowledge of the situation in any of them. None learned to trust the knower over the guesser. (By the way, children cross this comprehension threshold between the ages of three and four, but with a difference: the successful chimps required over 100 trials before understanding the problem; when kids are finally ready to 'get it', they do so fast, with just a dozen trials.)

In a similar experiment, the chimpanzee Sarah was shown a series of videotapes showing one of her trainers experiencing some kind of problem, such as trying to operate a radio which was clearly unplugged. Then the videotape was stopped before the problem was solved and Sarah was presented with a choice of pictures, one of which showed the correct solution to the problem. Sarah selected that correct answer at well above a chance rate. Interpretation: she understood a problem was being confronted and inferred a sense of purpose to the individual. But the really fascinating point here is that Sarah would choose incorrect answers if the trainer depicted with the problem was not one of her favourites — intriguing behaviour open to a range of interpretations, one of which is that Sarah was lying.

So Sarah does seem to have a concept of other minds, although this concept is limited, apparently. Other tests have suggested that she is not capable of understanding when the other mind holds a false belief. The set-up of this study was simple. A cabinet contained good food on one side and horrible, disgusting stuff on the other side. Sarah had no problem directing a trainer to the good food. Then a masked man entered the room and switched everything around. Sarah reacted with vehemence and threw things at this evil intruder. There's no question that she knew what had happened, but when the unknowing trainer — the trainer who held a now-false belief — came back in, Sarah did not seem to understand that he might not know about the treachery. She didn't try to warn him, and she directed him to the side of the cabinet that now held the bad food.

chapter 12

CONSCIOUSNESS, CONTINUED

Deception

We humans are not the only Machiavellians. Animals are good at politics and deception as well. R.W. Byrne and Andrew Whiten wrote a book on the subject entitled *Machiavellian Intelligence*, and every other book on animal behaviour has a chapter or two on the subject. Including this one, because deceptive activity is not only fascinating in its own right, but also provocative as it relates to the question of consciousness. Deception seems to require a sense of the other mind as well as one's own mind.

Let's begin this survey with the firefly, a common beetle whose overall communication skills are 'fantastically complex', in the words of Donald Griffin – a reference, in part, to the biochemistry of the luminescent flashing. The male of the species *Photinus pyralis* seeks out females on a summer's eve by flying near likely bushes while 'flashing its light' every six seconds for half a second. A responsive female in one of these bushes usually waits a couple of seconds before answering the male in the affirmative with her own half-second flash. The male then turns and flies in her direction, and increasingly excited flashes go back and forth between the pair. The 'excitement' part of the narrative is a blatantly anthropomorphic assumption on my part, but, nevertheless, mating usually follows in short order.

But there is a catch. It happens that the female of a larger species of firefly, *Photuris pyralis*, mimics the reply of the *Photinus* female. If the hapless *Photinus* male responds to this deception and comes too close, this deceptive female will eat him on the spot and thus get a quick and easy, high-protein meal for her own eggs. However, only about 10 to 15 per cent of the males that initially approach the deceptive *Photuris* female go all the way, so to speak, and put themselves in harm's way. Most somehow catch on to her ruse and fly away at warp speed!

Can the *Photuris femme fatale* really have an intention to deceive the *Photinus* male? And does this male actually understand that he has been deceived when he takes evasive action? We have to be cautious about drawing this conclusion, and the same is true, I believe, when considering some species of scorpionfly, in which the male must present the female with a gift of food – an insect – before she will mate with him. Males either do their own hunting for this purpose or, sometimes, present themselves as females to other males who already have their insect. They then steal this bug and head for the real females themselves. Intentional behaviour, or sheer instinct? Deception among insects is quite common. Perhaps they have to fool everyone because most of them are so small?

What about the behaviour of the genus *Gonodactylus*? These crea-tures are predatory by nature and operate from caves. The battle for a good cave often requires ritualised posturing and gesturing and, to all appearances, bluffing. They recognise other individuals – by way of odour, it is believed – and avoid the caves of those who have defeated them. Members of *Gonodactylus* are also known among specialists as 'stomatopods'. Among the rest of us they are known as shrimp – mantis shrimp, to be exact.

The individual whose research brought to light the surprisingly complex behaviour of this shrimp denies any consciousness for his subjects, as do most readers of this story, I suspect. They are shrimp, after all. But, as Griffin points out (I was very surprised to learn this), mantis shrimp happen to have a well-organised central nervous system, which is larger than a bee's. And we have seen what a bee can do. Therefore, Griffin concludes, '. . . it is quite reasonable to speculate that mantis shrimps may experience very simple conscious feelings or thoughts about the fights by which they gain or lose the cavities that are so important to them and the antagonists whose odors they learn to recognize from previous encounters.'

We already know that birds are quite intelligent in some respects, and that the brains of some birds are larger, relative to body weight, than the brains of the 'lower' mammals. Donald Griffin reports the following case of avian deception on the part of a hummingbird who had detected a mist net strung across his territory. (The episode was observed by researcher A.C. Kamil.) Everyone studying hummingbirds knows that, once a net has been discovered by a bird, it will never be effective against that bird, who knows it has to avoid it. This particular hummingbird even used this net as a perch! In fact, he happened to be perched on it when another male entered his territory. As we have learned, male hummingbirds are ferociously aggressive in defending their territories, usually charging directly at the intruder and driving him away. In this case, however, the net-knowledgeable defender of the territory flew *around* the perimeter of his acreage until the intruding bird was *between* him and the net. Then he attacked, successfully driving the unfortunate bird into the net.

Kamil does not relate this 'anecdote' for the purpose of attributing consciousness, but Griffin argues that it is suggestive of 'intentional planning', if nothing else. Even more suggestive of conscious planning and deception is the behaviour of the shrike. I can only imagine the weeks and months of careful observation that must have been necessary to establish this behaviour, now well-known and reported in many sources, including *The Animal Mind*. Briefly put, certain shrikes who dwell in the treetops serve as sentinels for all the birds feeding below them in the lower reaches of the trees. The shrikes watch for dangerous hawks, which will take smaller, unsuspecting birds on the wing. This deal works both ways, however, because the shrikes feed on insects stirred up by all the other birds below them.

As the shrikes move through the treetops, the other birds follow. When the shrike sees a hawk, it emits an alarm call and the other birds dive for cover. But sometimes, when a shrike sees another bird about to capture a particularly juicy insect, the shrike emits a false alarm; the other bird forgets the insect and flies for cover. The shrike then takes the meal. Should the other bird catch this insect despite the alarm call, the shrike may stifle the alarm mid-call. This behaviour seems pretty intentional to me, as though the shrike understands that he should not 'waste' alarm calls or the other birds will catch on to the trick.

The shrike is not the only bird which has been observed emitting false alarms in order to get rid of competition temporarily. In fact, many birds are capable of all sorts of deception, none more intriguing than the 'tonic immobility' of many ground-dwelling birds if they are captured by a predator. They play dead, in other words. (Why should it matter to the predator whether its prey is dead now or dead in a few minutes? The answer is unclear, but apparently it does matter, because sometimes the too-sudden lack of movement by the prey will induce the predator to drop the meal and move on. And as we all know, the recommended procedure for surviving a bear attack is to curl up and lie as still as possible, despite mauling.)

We assume that much of this feigning behaviour by birds is instinctive, and surely it is, but there are also good reasons for seeing a measure of versatility. It turns out (according to the results of clever, controlled experiments) that death-feigning birds, such as bobwhite quail, are more likely to play dead if they see two eyes in front of them, not just one. Why should this be? With very few exceptions, the predators of the world have a pair of eyes mounted fairly close together and looking forward. This allows for binocular vision, which in turns provides vastly superior information on distance, as we can all prove by trying to touch an object with one eye closed. On the other hand prey animals, such as wildebeest, have eyes mounted on either side of the head. This sacrifices 'distance control' but provides a wider range of vision, which is more important for them. Presumably, the bird responds to two eyes rather than just one because two eyes mean 'predator'. If the feigning bird discovers just one eye staring down, it is more likely to take flight, instead of going to the trouble of playing dead.

Perhaps the most famous deceptive behaviour in the animal kingdom is that of the female plover. This clever bird is world-famous – thanks to the exhaustive research of Carolyn Ristau of Rockefeller University, who contributed a lengthy essay on the piping plover to

Cognitive Ethology. This bird has at least four quite distinct tricks she employs to lure a predator away from her nest of eggs, including the 'broken wing' ruse, in which she presents herself to the predator in just such a vulnerable state, executes a brief run away from the nest, feigns again, then takes off for another brief run away from the nest. Another ploy is to run squeaking along the ground in the manner of a small rodent, which is also choice prey. A third is to sit in brood on a nest that is not there, always at some distance from the real nest. A fourth trick is to create a general and highly visible rumpus, tempting the predator, then retreating, tempting, retreating, tempting, retreating – always away from the nest, of course.

These deceptions are not restricted to birds. The hognose snake is justifiably famous among biologists for its extravagant display of horrible death, beginning with violent writhing, concluding with inert, upside-down stillness and bleeding from the mouth. Meanwhile, it remains quite alert to the behaviour of the predator. What to make of all these clever stratagems? The traditional explanation, before Ristau's research, was that most if not all such tricks are really just a classic case of 'conflict behaviour'. This much-studied phenomenon among animals results from a conflict between the 'fight' and 'flight' instincts and yields confused and pointless behaviour. (In tight situations, we humans are just as susceptible to conflict behaviour as any other creature.) However, it is very difficult to reconcile this explanation with the plover's behaviour, which gives no sense at all of being confused.

Sometimes the deceptive behaviour seems incorrect, in the sense of instinct gone awry. For example, the deception may continue after it's too late: the eggs have already been eaten. This doesn't seem very intelligent, and it's the kind of point that doubters pick up on with a vengeance. One researcher wrote, 'It is ludicrous to suppose that injury-simulation arose through birds deciding the trick was worth trying.'

Ristau does not deny that there must be a strong instinctive component in the behaviour, but she takes great care to point out the aspects of the behaviour that seem more thoughtful. The bird keeps an eye on the predator and adjusts its behaviour accordingly, in many cases. The action almost always leads away from the nest, a strategy that is not necessarily easy, as the 'engagement' shifts ground over time.

Also, the choice of deceptive trick seems to vary with the nature of the intruder, with the full-blown behaviour saved for the most dangerous. A mere cow is not a predator, so the plover doesn't bother with the broken wing display with a cow. However, the lumbering bovine could unwittingly tread on the nest with tragic results, so the plover does park herself between the approaching cow and the nest and make a racket, thereby usually diverting the animal. Most impressively, Ristau watched as the birds apparently discriminated between specific trespassers, based on their previous experience with these trespassers. And they drew their inferences after just one encounter. In order to ensure control, Ristau chose humans as the only suitable 'predator' for this phase of the experiment. Plovers dealt differently with the individual who sauntered near the nest but showed no interest as opposed to the individual who made what could be – and were – interpreted as threatening gestures toward the nest.

Ristau is a careful scientist. She concludes, in *Cognitive Ethology*, 'These experiments are only a beginning in the exploration whether . . . plovers are intentional creatures. The results so far suggest that they are.'

If deception might be a marker that an animal has a concept of other minds and its own mind, we should expect to find choice examples among predatory mammals and primates. And we do. The following story is told by Donald Griffin in *Animal Minds*. The narrative is based on Griffin's own experience at Amboseli National Park in Kenya while travelling with Robert Seyfarth and Dorothy Cheney, whose book about vervet monkeys I have cited quite a few times. Now, however, we are dealing with lions, not monkeys. One day Griffin and the Seyfarths were driving along a dirt road (there is no other kind in Amboseli) and paused to watch a herd of wildebeest, some on one side of the road, some on the other. While the trio watched, at least four, maybe five, lionesses approached with a 'businesslike gait'. (There is nothing more 'businesslike' than the gait of a hungry and confident contingent of lionesses.) All the wildebeest stopped feeding to observe them, and, without pausing, two of the lionesses forthrightly climbed a small knoll and sat there in clear view of all the wildebeest. They were not trying to hide; just the opposite, in fact, or so it seemed. Shortly thereafter, a third lioness slunk along the road in a ditch, belly to the ground, and moved into position between the two groups of wildebeest. This animal was trying to hide, obviously, and had succeeded.

All was quiet for a couple of minutes. Suddenly a fourth lioness charged towards the wildebeest from a copse of trees on one side of the road. The wildebeest on that side rushed to join the others on the other side of the road, but in order to get there they had to get past the lioness slinking down the ditch. She now sprang into action and took down one of the wildebeest. Soon the charging lioness, who had presumably been part of the group to begin with, and the two lionesses who had positioned themselves so conspicuously, sauntered over to the scene and commenced dining. The other wildebeest stopped to watch from 100 metres away.

The lions' behaviour in this story demonstrates mapping of the terrain, cooperative hunting, shrewd deception – a veritable galaxy of ingenious behaviours. So far as we know, lions are not capable of solving the reasoning problems in the laboratory that chimps can solve, but scenes like the one Griffin describes demonstrate again that we see animals at their most intelligent and most 'conscious' while they are engaged in the most pressing activities of life in the wild. You could observe captive lions for a lifetime and never see such intelligent behaviour.

Let's think for a moment about hiding. It seems to require a sense of self – otherwise, why hide? – but the range of animals that hide is enormous, indeed. Snakes do it. Invertebrates. Birds. Mammals, of course. Many reports have come in regarding grizzly bears – always grizzlies, it seems – who position themselves in vantage points from which they can see but not be seen, and Donald Griffin reports one story about the grizzly who hid its own tracks. Now this, if confirmed, would say a great deal indeed! Or is such behaviour qualitatively different from a cat's instinctive efforts to cover its tracks, so to speak, after it has finished its business in a litter box? Anyway, practically everybody out there hides, and if we stop to consider the phenomenon, it can often be explained as a simple process of learning: the last time I got into this dark place I walked away alive, while my friend who stayed outside got eaten. So hiding may be a clever form of deception, but it is not evidence that an animal understands 'I am hiding.'

Among the primates, however, apparently purposive deception is everywhere. You could almost say that it's a way of life. A common and well-substantiated trick is for one male to 'ignore' an object – food, very likely – while in the presence of an alpha male, or any higher-ranking animal, in the hope of having unchallenged access to this object later on. A widely reported example of this evasive behaviour is that of a male mangabey, a close relative of the notoriously fractious baboon, who was given a sneak preview of food before his entire group was released into an enclosure. The first time this happened, this male went straight for the food but then had it taken away by a dominant male. In the second trial, the male stayed away from the food and was followed by the dominant animal, and both lost out to an enterprising third party. In the third trial, the subject male headed away from the food, bided his time, then rushed for, and gained, the prize. In all following trials, he followed this same procedure with good success.

Frans de Waal observed one ape wait several hours until the coast was clear, an impressive demonstration of deferred gratification. (It is quite difficult even for a chimpanzee, and almost impossible for other animals, to voluntarily suppress the desire for the immediate gratification of food.)

Marc Hauser reports that the macaques he observes on the Puerto Rican island of Cayo de Santiago ('Monkey Island') routinely assess the social situation before calling out about a new food source. Who's around? Friend or foe? Kin or non-kin? These are critical concerns, because if the monkey lies, in effect, by not reporting the food with foes and non-kin around, they will rough him up. What's most fascinating to Hauser is the behaviour of the 'peripheral male' who isn't a member of any group. This male almost never makes a food announcement and, even when caught, is never beaten up. This suggests that the rules of engagement, so to speak, are specific to the one social group.

Writing in *Bonobo: The Forgotten Ape*, de Waal tells us about the infant chimp Laura who, when instructed to clean her plate, did so amazingly quickly. Her caretakers were unable to work out how she had accomplished this until they discovered the food in her nappy. Several times de Waal observed one of his primate friends making overtures of reconciliation as a ploy to lure the enemy within reach, then attacking.

One final story from his book:

> A juvenile male tentatively reached for an apple that had rolled away from his mom's food pile. He stopped, seeming to realize that picking up the fruit would get him into trouble. Instead, he approached his younger sister, with an exaggerated play face. She was sitting near her mother in the direction of the apple but had not noticed it. While wrestling with his sister, the male got closer and closer to the apple. Then, with a sudden movement, he grabbed it. His interest in the play vanished . . .

In Wolfgang Kohler's famous dangling-banana experiments in the Canary Islands, so important in the study of tool use among the apes, dominant males often allowed their inferiors to build the tower of boxes that would reach the banana, then took over at the last moment – or tried to take over, because the inferiors often managed to destroy the tower in the process of vacating the scene. Accidental or intentional destruction?

An episode of deception which 'made the rounds' – and therefore the literature – among ethologists was the behaviour of the female baboon who sneaked away from the alpha male in her troop and mated with a young male behind a rock, unseen by the alpha. Moreover, she peeked out from behind the rock, apparently to make certain that the main male remained oblivious to her infidelity. Similar behaviour has been observed among the macaques on Cayo de Santiago. There have also been many reports of primates engaging in 'acoustic hiding', as it's called – repressing the usual sounds of sexual behaviour if this behaviour might run afoul of the established order.

Jane Goodall tells us about the male chimpanzee, Humphrey, who had the nasty habit of attacking females quite brutally when greeting them. One day Fifi sees Humphrey coming, and he's obviously in one of his bad moods. She looks around nervously, then suddenly stops and stares into the nearby bushes and gives a little hoot. Humphrey, 'all bristling and swaggering', as Goodall describes him, strides to the bushes to investigate and conquer. But there's nothing there, and by the time he returns to Fifi, he has calmed down and grooms her instead of attacking. And Goodall has the incredible story of how chimpanzees are able to maintain almost total silence while out on patrol. They avoid dry leaves and rustling bushes. One time they maintained vocal silence for three hours: acoustic deception of a very high and calculated order.

Imitation

At the Camp Leakey Research Centre in Borneo an adult orangutan decides to follow suit as one of the local staff washes his clothes and takes a sponge bath on a pier over the lake. When the young man washes, the orangutan washes; when the man squeezes his clothes, the great ape squeezes the towel it has brought along; when the man pours water over his head, his shadow pours water over his head. There is no question that the orangutan is intrigued by the human's behaviour and copying it gesture for gesture. It's a charming scene, and also an important one, because cognitive ethologists consider the ability to imitate another's behaviour to be a sophisticated cognitive and social skill.

Imitation is the basis of cultural learning, which is one of the most powerful kinds of learning. Moreover, imitation, when broken down into its components, strongly suggests that the animal is able to keep an image or action from its environment 'in mind' and then transfer this image or action to its own body. It suggests a sense of self – one that infants pick up quite early, at around one year old, according to Piaget. (In some experiments, babies less than a month old mimicked facial expressions such as sticking out the tongue – or is this just 'deferred motor response', as Jacques Vauclair asks, rather than the 'mental reconstruction of an entire motor sequence'? The question remains open, I believe.)

Imitation is an incredibly powerful, and quick, form of learning. Rather than having to learn everything through your own mistakes, you can learn by watching others' successes. Jane Goodall and Christophe Boesch have each observed chimpanzees learning to 'fish' for ants, one in Gombe, East Africa, and the other in the Tai forest in West Africa. In both cases young chimps study their mothers' actions intently before stealing or picking their own grass wand to try it for themselves. But when Goodall and Boesch compared notes they realised that the two groups of chimps were using their fishing tools in very different ways. East African chimps use a long wand to gather as many termites as possible, sweeping them off with their hand and gobbling them down. Whereas chimps from West Africa use a much shorter stick, which picks up far fewer termites, picked off with their lips. These methods are passed down from mother to offspring across generation, resulting in two different 'tool cultures'.

But animal 'culture' isn't always what it seems. Remember the cheeky blue tits that learned to peel off milk-bottle tops to drink the cream at the top? The behaviour spread slowly across the English blue tit population (much to the horror of affected householders!) suggesting that the blue tits must be watching their neighbours stealing milk and learning by imitation, copying their behaviour. But studies with captive birds showed that something much simpler was afoot – the birds were acquiring this skill by a form of simple trial-and-error learning called 'stimulus enhancement'. A blue tit watching another bird peeling back a milk bottle lid didn't actually *understand* what the other bird was doing. It didn't relate the other bird's behaviour to its own and think 'I could do that too', it merely got interested in the *milk bottle*. So next time the watching bird saw an opened milk bottle it went to investigate, and, following its instinct to probe around, discovered a rich source of milk! And so what looks like imitation actually turned out to be a form of roused curiosity (or stimulus enhancement, as scientists like to call it).

The most famous case of imitation among animals is probably the potato-washing habit of a troop of macaques living on the Japanese island of Koshima. Or was it imitation? In the early 1950s, observers noticed that one female, Imo, carried her sweet potato to the surf and washed it. (This island was a preserve and the home of Moza the snow monkey. Sweet potatoes and wheat were periodically dumped on the beach.) Very slowly the washing habit spread through most of the troop. Three years after Imo's breakthrough, 10 other macaques were washing their tubers (out of 60 animals); and after six years, 17 macaques were food-washers. Some years after Imo began washing her sweet potatoes, she began washing her wheat. The sand sank, leaving the grains floating in the surf, where she scooped them up. Clever, but again, her fellow macaques slowly picked up the habit, one by one. But is this true imitation? Did Imo's troupe members *understand* what Imo was doing and decide to follow suit? Or can simple stimulus enhancement explain the behaviour? The jury is still out, but, given the long time it took for the habit to pass from one macaque to another, many psychologists are betting on increased curiosity combined with trial-and-error learning.

So where does that leave imitation? Can all the examples we've seen be explained away in simple terms? As always, sometimes the only way to really find out what's going on in the animal mind is to give them a test.

Professor Andrew Whiten, of St Andrew's University in Scotland, has been working with Sue Savage-Rumbaugh and her chimpanzees to see if chimps really do learn by imitation or whether it's all just simple trial and error. Whiten devised a complicated puzzle box, which he calls an 'artificial fruit' because it's complicated to get into (like many fruits in the tropical forest) but contains something delicious. The opening mechanism involves two bolts (that can be either twisted or poked out), a pin (that can be twisted or spun out) and a handle (that can be turned or pulled). It may sound a bit like a Chinese puzzle box but Whiten makes it easier for the chimpanzee by showing her how to open the box. In fact he shows her an exact sequence (bolt twisted, pin spun and handle turned) three times, and then lets the chimp have a go.

Now this is an excellent test because, depending on how the chimp opens the box, we can see if she's learning by trial and error or by imitation. The box can actually be opened in several different ways, which are in fact easier than the way Whiten showed her — so if she learns by trial and error she will open the box any old way. But if she learns by imitation, she will follow Whiten's exact sequence of behaviours. And guess what? The chimp copies Whiten almost exactly.

This is fairly conclusive proof that chimps *can* imitate. And that they're capable of true culture. But what of other large-brained mammals, like dolphins? We've often seen them at oceanariums doing tricks and imitating trainers. But is this true imitation? Do they have a true understanding of the aims and actions of others and are they able to relate this to themselves?

In Hawaii, psychologist and dolphin researcher Lou Herman has been working with his two beloved dolphins, Ake'kamai and Phoenix, for nearly 20 years. He had observed dolphins in the wild imitating a wide variety of behaviour by human divers, seals, turtles and skates, and so decided to devise a test to see if Ake'kamai and Phoenix could imitate — by finding out if they could copy an action the first time they were shown it (so there would be no time for trial-and-error learning).

The set up is this. Lou asks Ake'kamai to do a back-flip. Phoenix can't see the instruction, but can see Ake'kamai leaping up into the air, twirling round and splashing back down to the water. Herman then asks Phoenix to imitate what she's just seen and she copies perfectly, rising up above the water with a flip of her strong tail. The next time Ake'kamai is asked to do a balletic underwater somersault,

Phoenix sees only the behaviour before being asked to imitate, which she does to a fault, first time. No room for trial and error there.

And not only do they seem to have an understanding of another's actions, they are able to remember and repeat their *own* actions. Once Phoenix has done a back-flip you can ask her to repeat what she's just done and she will. Such tests might seem basic, but there's beauty in simplicity and Herman's work tells us a lot about the dolphin mind. Not only is Phoenix aware of another dolphin's actions and able to relate their body parts to her own, she's also aware of her own actions. How can dolphins follow these instructions without having a sense of what they've just done, and how would they have this sense without having a sense of the 'I' who's just done something?

The Question of *Cephalopods*

Perhaps no creature poses such provocative questions about imitation and self-awareness as the *Coleoid cephalopods* – the squid, the cuttlefish, and the octopus. Graziano Fiorito reports, in *Science*, on experiments with octopuses in which one observer octopus watches as another learns the rule in a red ball/white ball game. The rule is simple –attack the red ball, ignore the white ball – and the first subject learns it quickly. Given his chance, the observer octopus performs almost flawlessly almost immediately. But these experiments have never been repeated, and the octopus's imitation ability has not been investigated as thoroughly as that of the chimps or dolphins.

Writing in *Anthropomorphism, Anecdotes, and Animals*, Martin Moynihan provides a report on the amazing mimicry among *Sepioteuthis sepioidea* – the Caribbean Reef Squid, the most studied of these creatures: 'Two young [squids] were discovered in very shallow water over white sand at midday . . . Since they could not easily escape from the observer, they first shot out blobs of ink and then turned dark and floated with twisted arms, in obvious imitation of the blobs which were slowly disintegrating. The animals were, in fact, mimicking their own decoys.' And this sort of display is just for starters.

These squids can change colour to an almost infinite degree, thanks to special pigment cells in their very thin skin. Each of these cells, with its supply of special chemicals, is individually hooked into the animal's central nervous system and can be manipulated independently of all the others. Changes are almost instantaneous, and result in a light show that virtually defies belief. Reef squids create mesmerising waves and patterns of colour in mating rituals, and equally amazing but dissimilar patterns when faced with a dangerous predator. They camouflage themselves – and not just by turning black when in front of a black object, or white with a white object. They can camouflage themselves when positioned against a half-white, half-black object, and the line between the two colours in the squid is clean and sharp. Moynihan writes, 'The arrangement is never random. It is precisely the parts of the body that are near white that go white. The same for black. The parts may be left, right, front, rear, top, and/or bottom. All combinations are possible. One part may be black during one performance, white during the next, depending upon circumstances. More complex backgrounds may encourage more complex patterns of color.'

When a group of Caribbean squids is confronted by a mildly threatening predator such as a Yellow Jack, most of the group will immediately project the same pattern, usually either 'bars' or 'streaks'. But often the last squid to change will go exactly the opposite of all the others, and thus intentionally stand out in the eyes of the predator. What could be the point? Moynihan does not know, but it may well be a wilful act. The truth is, we have no idea what communication is transpiring in the complicated exchanges observed between these squids. They are, as Moynihan writes, 'kaleidoscopic and baffling . . . There must be a switchboard linked to something like a cognitive map.'

Moynihan was a senior scientist at the Smithsonian Tropical Research Institute in Panama who wrote about primates as well as these cephalopods, and he was convinced that 'awareness' should be our 'default' assumption when it comes to complex behaviour in animals. He says, concluding his fascinating discussion, 'The flexibility and opportunism of *coleoid* behaviour is suggestive. These animals have considerable abilities to solve different social problems in different circumstances. Choices are overt and decisive. I cannot believe that they are not deliberate and, in some sense, conscious. Alternative explanations would seem to be even more complicated, with even more loops and feedbacks . . .' A reminder of Morgan's Canon. Sometimes the parsimonious, or simplest, explanation for a behaviour is that the animal is conscious and self-conscious!

The Mirror on the Wall

The line between consciousness and self-consciousness may be a fine one indeed – it might be the proverbial distinction without a difference – but self-consciousness (awareness of myself) is generally considered the highest order of consciousness that we *Homo sapiens* know anything about. Until now, I have addressed this question of self-consciousness obliquely and by inference, and of course all explanations have been challenged. Understanding the limitations of inference, psychologists have devised tests to seek out self-consciousness directly.

The tool employed is the mirror. Sounds good: we humans look into mirrors explicitly to see ourselves, and we understand that the image we see is ourselves. Do animals? With some primates, this would certainly seem possible, because observers have reported for many decades that chimps stare at their reflected image in still water, and even reach out to touch that reflection. Wolfgang Kohler in the Canary Islands was the first serious researcher to report that primates could recognise themselves in the mirror, and the systematic investigation of this behaviour was started by Gordon Gallup, a psychologist at the State University of New York at Albany. Today, mirror self-recognition, or MSR, tests are a staple of comparative psychology, and they require our close attention.

In the classic test, the subject animal – a chimpanzee, say – is introduced to a mirror over a period of time. (Chimps have no problem understanding the basic idea of a mirror.) When the animal is thoroughly familiar with the concept, she is then anaesthetised and a small mark of some kind is put on her forehead. How will the chimp react when she later sees this mark in the mirror? Will she recognise that it is a mark on herself? In most of the early studies, the chimp almost immediately used the mirror to investigate the mark on her forehead. She went straight to it with a finger. She knew that the image in the mirror was herself. Thereafter, the chimp was usually even more interested in mirrors than she had been before.

Children pass the MSR test at about the age of two, when they also begin to show the self-conscious emotions, shame and embarrassment. For many years, the test was considered proof positive of self-consciousness on the part of chimpanzees. Who could doubt the result? How could it be interpreted otherwise?

Well, Celia Heyes can interpret the test otherwise. We know Heyes by now as one who takes a very conservative position on issues of animal thinking and consciousness, and she does not disappoint us in this case. She believes the mirror test can be explained as using visual feedback at an advanced level, but not as indicating a true sense of self. An animal does not have to have a sense of its own uniqueness in order to pass the test, she argues. Just as a chimpanzee could look through a telescope at the stars and not see the stars as we see them, so she can look into the mirror and not see a unique creature there. In short, self-recognition is not necessarily self-consciousness, just as 'cognition' is not necessarily 'thinking'. Or take the stag who knows how wide its antlers are and doesn't attempt to walk through too small an opening, or the fact that many mammals know that their shadows are related to their bodies. Does either instance require true consciousness on the part of the animal? Not really, even according to Donald Griffin, who thinks of this as 'a sort of self-awareness'. The stag might have learned about narrow passages between trees from harsh experience, and the horse, say, has become habituated 'to the dark area on the ground that accompanies it on sunny days'.

Heyes is not alone in her negative assessment of the MSR test. Nicholas Humphrey, who popularised the idea that certain animals – chimps among them – are natural psychologists, did not believe that primates are self-conscious. In the 1980s, one group of careful and not unsympathetic scientists ran some MSR tests and found much less 'mark-directed behaviour' than they had expected. In one of their studies involving 11 chimps, a majority showed interest in the mirror and three showed other 'self-directed' behaviour following dozens of hours of exposure, but only one officially passed the MSR test. (This chimp understood almost immediately.) These results were very puzzling, and in fact the authors had trouble getting their work published because it ran so contrary to the presumption in primatology about the 'robustness' of MSR among chimpanzees.

The deeper the investigation of animals and mirrors, the more puzzling the results. No elephant has ever passed the MSR test. Dogs do not pass the mark test, but dogs do engage mirrors in interesting ways, as any dog owner can prove to his or her own satisfaction, with a great deal of patience. One trial seemed to demonstrate that pigeons can pass the MSR test, but these results have been noisily challenged, as we might imagine. And yet the use of mirrors as visual feedback is surprisingly widespread among animals who are occasionally studied in the laboratory.

Sue Savage-Rumbaugh's bonobo Kanzi seemed to like watching himself blow bubbles in the mirror, and the chimpanzee Austin enjoyed watching himself demonstrate the astonishingly adroit lips of all chimps by rolling a dime on edge from one side of his mouth to the other, and then back. Chimps can use television images to guide them in their manipulation of objects, and they can use the image to guide their arm while reaching for a banana – 'a decidedly nontrivial bit of self-recognition', in the view of Daniel Dennett, who's no pushover on this question. There is a debate as to whether gorillas pass the test. They definitely have trouble with the standard set-up, but Penny Patterson's famous Koko uses a mirror to groom, and recent studies have shown 'self-directed' and 'mark-directed' behaviour.

It turns out that the whole issue of mirror self-recognition is a classic case, not of anthropomorphism but of anthropocentrism. We are looking for proof of our kind of self-awareness because this is the only kind we ourselves can understand. Any others are off limits, literally. Monkeys are an excellent example. Confronted with a mirror image, monkeys see another monkey, judged by their vocalisations and, sometimes, threatening gestures. Allowed to handle a mirror, most baboons and monkeys will try to find the animal behind it, or try to kiss or lick the image. They do not seem to recognise the image for what it is: themselves. Or is this the behaviour of young monkeys? In some cases older monkeys do not try to find out what's behind the mirror and sit staring at it, although without interaction. And they may look behind them when they see someone in a mirror. So they know what a mirror is all about. Furthermore, when toys are attached to their necks, monkeys will use the mirror to engage the toys.

These results would seem to indicate that the monkeys understand just about everything regarding mirrors except the image in front of them. This is possible. Or it is also possible, as Sue Savage-Rumbaugh concludes after her own analysis of the MSR test, that monkeys are just not interested in the dot on their heads. That's why they don't react when they see it in the mirror. This is a fascinating hypothesis, and it gains some credence when we learn from Savage-Rumbaugh that chimps and baboons will clean themselves of any 'sticky-gooey fruit', while monkeys will not bother. She writes, 'Perhaps monkeys are more dependent on other monkeys for self-tidiness than apes, or perhaps their self-concept does not deal as much with appearance as does the self-concept of apes.' Other commentators have pointed out that monkeys and gorillas might not do well on MSR tests because of cultural constraints against looking other animals in the eye, which is taken by them as an act of bold aggression. On film we see the bonobo Panbanisha 'pass' the MSR test with nonchalance, casually removing the dot that had been surreptitiously placed on her forehead. Apparently, she did not know she was taking an internationally televised test!

Dolphins offer another excellent illustration of the problem. Ken Martin, director of Earthtrust at Hawaii's Sea Life Park, who has tested dolphins with the mirror, says:

> You see them kind of play with their image, which is called contingency checking. They move their pectoral fin, or their head, or they open their mouth and watch the contingency of that in the mirror. It's like my seeing myself on television and waving my hand. That's one of the ways you recognise it's yourself. Work has been done with killer whales who take that to an extreme. They do rhythmic bubble blowing, and open their mouths, and stick out their tongues.

Since dolphins and whales can't touch a mark on the forehead, the classic MSR test is not possible with them. One alternative method Martin has come up with is to carefully compare mirror behaviour with normal social behaviour. And it turns out that dolphins are always very calm looking in the mirror, while they're much more active – hyperactive, even, from our perspective – when dealing with each other. Like everyone who works with intelligent social creatures, Martin believes the cetaceans need self-awareness and other-awareness in order to deal with the complex society they live in, although this is a very different kind of society from our own. Acoustic identification is much more important to dolphins than visual identification. (The large cerebral cortex of dolphins and other cetaceans seems to be somewhat differently organised from the cortex of primates, and not as structurally complex. Or is this just in the eye of the beholder, who happens to be a primate?)

Dolphin society is closer to the matriarchal society of elephants than to that of primates, including us. In fact, Martin thinks of primates as rugged individualists, and dolphins as very clever sheep. And he says:

> When you look at dolphins in the wild they are always with others, they do everything together, they do everything in

synchrony. Being a dolphin is very much a team-player concept. Some form of group awareness might be very important for dolphins. They have common ancestors with the hoofed mammals, and it could well be that group awareness might be at least as important to them as self-awareness. In fact, self-awareness might be a strange concept to dolphins. You never see one alone in the ocean. Never.

So what if these cetaceans can't pass our MSR test? Are we going to say — on these grounds alone — that they can't have self-consciousness? This is absurd. It may very well be that primates and cetaceans and other animals may be self-conscious on their own terms, which are, after all, the only ones that count to them.

chapter 13

THE HEART OF THE MATTER

They Can Suffer

Until this point I have avoided, for the most part, the issue of
suffering in animals, but now is the time to confront this question.
Daniel Dennett believes that the capacity for suffering may be an
important benchmark for consciousness. As he writes, in *Consciousness
Explained*, 'The capacity to suffer is a function of the capacity to have
articulated wide-ranging, highly discriminative desires, expectations,
and other sophisticated mental states.' I agree with Dennett, but I
agree with Jeremy Bentham more. Outraged by the arguments of
the Cartesians, who used our human rationality as the one and only
threshold for consciousness, this eighteenth-century English utilitarian
wrote, '. . . a full-grown horse or dog is beyond comparison a more
rational, as we all are a more conversable animal, than an infant of a
day or a week or even a month, old. But suppose they were otherwise,
what would it avail? The question is not, Can they reason? Or, Can
they talk? But, Can they suffer?'

Yes. All the aspects of animal minds and consciousness we have
examined in this book are fascinating in and of themselves – we
humans are curious creatures; we want to know things – but
the crucial point is that they influence far-reaching questions of
politics and ethics. This is a short chapter, but it may be the most
important.

Do we feel guilt – maybe just a little, furtive twinge of conscience

– about our treatment of animals? Many of us, yes, on issues ranging from eating meat, to torturing animals, virtually, for purposes of experimentation. In his book *In the Company of Animals*, James Serpell suggests that this moral conflict stems from our reflexive, innate anthropomorphism. Otherwise, why would we care? Otherwise, why have we designated pets as a protected category of animals whom we treat exactly the opposite of those animals we eat or put into cages? In Western society, tens of millions of dogs and cats and other relatively small creatures eat and live better in all respects than do millions of people in the less-developed countries. But a paradox within a paradox stares us in the face: while most of us treat our own pets with tender loving care and fuss about the quality of their diets – many eat a healthier, more balanced diet than we do – in the United States and the United Kingdom we put millions of other dogs and cats to death every year because nobody wants them.

In order to address this double paradox, we have a choice, Serpell argues: we can either deal with the ethics of our cold-hearted treatment of pigs, which are intelligent creatures, or we can laugh at and disparage our tender-heartedness towards dogs, who are no cleverer than pigs. In Western society, at least, we choose the latter alternative. Our attitude has been, quoting Serpell: 'Humans have a right to eat meat; farmers have a duty to supply this demand as cheaply as possible and animals inevitably suffer as a consequence. Why complicate the issue with imponderable questions about the morality of it all?'

It is correct that in 'the old days' life used to be much more difficult than it is today, at least in the developed world. Hunting and husbandry and domestication were economically necessary in the new scheme of things introduced in the Neolithic Age. Animals were wealth, plain and simple. (The word 'pecuniary', formerly much more common than it is today, stems from the Latin *pecus*, cattle. The Indian rupee comes from the Hindu word *rupa*, one particular cow. Many ancient coins denominations were designated by a particular animal.)

In our new Information Age, far removed as it is in many

ways from that earlier one, husbandry and domestication are still economically necessary. Animals are still wealth of one form or another, although cultural norms play a large role in how the details work out, culture to culture. The usual exhibit in this regard is the cow: sacred in India, eaten with gusto here. Dogs in the Western world are practically sacred in a 'secular' sense. (Throughout *The Descent of Man* Darwin emphasised dogs as evidence of evolutionary continuity for the good reason that he knew his primary audience, the English, were inordinate dog-lovers one and all.) But dogs are just 'dogs' in much of the rest of the world, the house speciality of certain restaurants in Vietnam and Korea (including South Korea, whose busy ministers of trade undoubtedly hope that this quaint gustatory practice remains little-known on our side of the planet). Roger Caras, President of the American Society for the Prevention of Cruelty to Animals (ASPCA), suggests that dogs became pets on this side of the Pacific and on the other side of the Atlantic because they're not a good meal in terms of meat on the bone. Dogs are skinny compared with, say . . . pigs. So one is pampered and the other tenderised.

Peter Singer sweeps right past the cultural distinctions and makes a very convincing case that the exploitation of any animal is morally equal to the exploitation of any human. When it comes to the question of which animals suffer, he doesn't believe in drawing lines. As far as I know, it was Singer's book, *Animal Liberation*, first published in 1975, that initially brought the issue before a large audience in this generation. (*Animal Machines: The New Factory Farming Industry*, by Ruth Harrison, was published a decade earlier, but I don't think it had the same impact. All apologies if I'm wrong.)

Singer essentially draws a direct analogy between the reasons for inferring animal consciousness that we have looked at in this book and the reasons for inferring animal pain and suffering: equivalent nervous systems, observable behaviour, and the evolutionary usefulness of pain and suffering. Then he takes up the argument of Bentham and other utilitarians that the principle of equality among humans is not a description, but a prescription. Equality cannot be a description, obviously, because humans are not equal in any 'performance category' you care to consider. But equality can and must be a prescription because otherwise someone has the power to draw the line. Otherwise, you could find yourself defending the morality of experimentation on babies as being preferable to experimentation on adults on the grounds that babies are our intellectual inferiors. Or, as Thomas Jefferson stated, 'Because Sir Isaac Newton was superior to others in understanding, he was not therefore lord of the property or persons of others.' (And yet Jefferson was himself a slave-holder, albeit one with a very guilty conscience who feared the day when God would wreak his vengeance on this culture.)

In conclusion Singer asks, 'If possessing a higher degree of intelligence does not entitle one human to use another for his or her own ends, how can it entitle humans to exploit non-humans for the same ends?' The only option, Singer argues, is speciesism, an ungainly word for which he apologises but doesn't find a suitable substitute. It is strictly analogous to sexism and racism.

So, on what ethical grounds do we support marine theme parks, whose large-brained performing cetaceans are, in the final analysis, enslaved? How do we condone, or at least rationalise, experimental research during which animals undoubtedly suffer? The capacity for suffering – in itself – is necessary reason for us to take into account the well-being of these animals. The fact that so many species breed only with extreme difficulty in captivity – with such difficulty that I should perhaps use the passive voice 'can be bred' – seems to me and others a clear indication that such captivity is emotionally traumatic.

I would be remiss, however, if I did not take note of the vast improvements which have been made in many zoos, animal parks and marine life facilities during the past 25 years or so. In today's best zoos huge sums of money go into the design and construction of 'natural' habitats for the animals and some, such as the San Diego Zoo, have large open spaces where the larger animals can roam about. New York's Wildlife Conservation Society International (formerly the Bronx Zoo) has been a leader in habitat design and has a new gorilla facility which may be the finest in the world. Many defenders of zoos argue that some endangered species would soon become extinct were it not for zoos' captive breeding programs and this is undoubtedly true. Nevertheless, conditions in some small commercial zoos remain deplorable and most of them should be shut down by the authorities.

And then there is the question of animal experimentation, such as the experiments conducted by Harry Harlow concerning maternal-deprivation in monkeys. Unsurprising result: the baby monkeys suffer and become depressed. It was Jeffrey Masson's book *Dogs Never Lie About Love* that introduced me to the English philosopher Mary Midgley's devastating critique of these experiments, from an ethical perspective. Her point could not be simpler: if monkeys and humans are comparable emotionally then the cruelty of the work is self-evident; if monkeys and humans are not emotionally comparable, then what is the point? Other writers have made a slightly different, equally devastating case: if we use the results of animal experimentation to draw conclusions about human physiology (most obviously regarding its reaction to new medications), we are logically obligated to infer in the opposite direction, so to speak, that conditions painful in humans must also be painful in animals.

Years ago, Jane Goodall was dismissed as 'a *National Geographic* cover girl who didn't know anything'. (That's her phrase, by the way, and I believe we can safely read a little sarcasm into it.) Goodall's descriptions of the emotional capacities of the chimpanzees at her research centre in Tanzania, were dismissed as — what else — anthropomorphic and anecdotal. Today, Jane Goodall is the most famous primate researcher in the world. In her characteristically blunt fashion Goodall says:

> Certainly anyone who's had anything to do with primates or pigs really does know that they're intelligent. You can't spend time with animals and not realise this unless you're completely stupid yourself . . . And I think also if you deny animals intelligence, even when it's obviously there, it makes it very much easier to do some of the extraordinarily unpleasant things that so many people do to animals. It's more comfortable to think that if you're cutting a piece out

of a dog for some kind of animal experimentation, well, he's not intelligent, therefore he can't feel like we do.

For years, Jane had heard stories about animal labs but had avoided going inside one because she knew the experience would be utterly depressing and that there would be nothing she could do to help. Then, in 1987, she was home from Africa, on holiday with her family in Bournemouth, England. As she relates, in the essay she wrote in *Intimate Nature: The Bond Between Women and Animals*, the whole family watched a videotape of scenes from inside a biomedical research lab in which 'monkeys paced round and round, back and forth, within incredibly small cages stacked one on top of the other, and young chimpanzees, in similar tiny prisons, rocked back and forth or from side to side, far gone in misery and despair'.

That footage changed her attitude. Now Jane felt she had to see these places for herself, and do what she could do. The lab had no choice but to open its doors to her. She was, after all, Jane Goodall. The scene she encountered was a 'nightmare', worse even than she expected. On a subsequent visit to a different lab owned by the same company, she met JoJo, a chimp born in Africa and shipped to the United States ten years earlier, now living in a cage with an old rubber tyre and nothing else. Goodall writes, 'Very gently JoJo reached one great finger through the steel bars and touched one of the tears that slipped out above my mask, then went on grooming the back of my wrist. So gently. Ignoring the rattling of cages, the clank of steel on steel, the violent swaying of imprisoned bodies beating against the bars as the other male chimps greeted the veterinarian.'

She now introduces into her narrative Jim Mahoney, who had been doing what he could to improve conditions in this lab, which was studying hepatitis, AIDS, and other viral diseases, but his means were limited. 'His round over, Jim returned to where I still crouched before JoJo. The tears were falling faster now. "Jane, please don't," Jim said, squatting beside me and putting his arm around me. "Please don't. I have to face this every morning of my life."'

Jane asks whether chimps should be used in these experiments at all, no matter how valuable the resulting knowledge. Do we have the right? No, she believes, we do not. Some defenders of the industry have called her a rabid anti-vivisectionist, and she has been tempted to call them in turn sadistic vivisectionists. But she realises that no purpose is accomplished by name-calling. The key to closing these labs, she believes, is education regarding the mental and emotional attributes of the primates. She told me, 'Humans are a species capable of compassion, and we should develop a heightened moral responsibility for beings who are so like ourselves.'

When she saw JoJo she thought of David Greybeard, one of the chimps in Gombe who opened the door for her into that 'magic world'. In her essay she continues:

> I had been following David one day, struggling through dense undergrowth near a stream. I was thankful when he stopped to rest, and I sat near him. Close by I noticed the fallen red fruit of an oil nut palm, a favorite food of chimpanzees. I picked it up and held it out to David on the palm of my hand. For a moment I thought he would ignore my gesture. But then he took the nut, let it fall to the ground and, with the same movement, very gently closed his fingers around my hand. He glanced at my face, let go of my hand, then turned away. I understood his message: 'I don't want the nut, but it was nice of you to offer it.'

For JoJo and thousands of other primates in their cages, 'nature's sounds are gone, the sounds of running water, of wind in the branches, of chimpanzee calls that ring out so clear and rise up through the treetops and drift away in the hills. The comforts are gone, the soft leafy floor of the forest, the springy branches from which sleeping nests can be made. All are gone.'

Can there be any question that our biblically ordained as well as self-proclaimed hegemony over animals comes with – or should come with – great responsibilities? The first national recognition of this fact was a bill introduced in the House of Commons in 1821 and promptly laughed out of consideration. Just two years later, however, the Society for the Prevention of Cruelty to Animals was founded, and then received a Royal Charter in 1840.

Today, certain countries, notably England, Sweden and Switzerland, have meaningfully tough, though not perfect, anti-vivisection laws. Many more nations have no laws whatsoever. There are none in the United States. The Animal Welfare Act of 1970, last amended in 1985, is woefully insufficient. Fully 90 per cent of all laboratory animals – rats, mice and birds – are completely exempt from all provisions, including those concerning housing and transportation. But the biggest loophole states that nothing in the law shall interfere with the 'experimental design' of a testing programme. In other words, if an integral part of the experimental design requires starving the subject animals, starvation is fine. Medical experimentation on brain-dead humans is illegal, while any equivalent experimentation on fully conscious chimpanzees might be completely legal. Furthermore, the review boards established to oversee experimental designs are staffed by the institutions conducting the research. As is so often true in the United States and other countries, the fox is guarding the hen coop.

Cruelty to animals by individuals is covered by individual American states but only 22 of the 50 states have felony laws on the books. In New York, where I live, for example, it may be a felony to set my neighbour's house on fire, but it is not a felony to douse his dog with petrol and toss a lighted match. The section of the New York statute addressing experimentation is also weak and full of loopholes. For example, one section reads, 'Pain and discomfort shall be minimized by proper use of tranquilizers, analgesics, and anesthetics. *Exceptions may be made when provisions . . . would defeat the purpose of the experiment.*' (My emphasis.)

On the other hand some ethicists as well as laypeople are afraid that this question cuts both ways. As I said in my Preface, they are afraid that if we do not maintain a hard-and-fast distinction between ourselves and all other animals we might find ourselves on the edge of a slippery slope that could lead to all manner of abuses of humans – the same abuses we now subject other creatures to, on the grounds that they are fundamentally different from us. If they are not fundamentally different from us, we are not fundamentally different from them. What's good for the goose . . .

Donald Griffin replies, in *Animal Minds*, 'Morals and ethics should surely be based on accurate understanding of the relevant facts, and since they have survived the Copernican and Darwinian revolutions, strengthened rather than weakened by correction of factual errors, there is no reason to fear a different outcome once evolutionary continuity of mentality is recognized.'

Questions about animal minds are not merely fascinating. The answers matter for our treatment of animals and, perhaps, of ourselves as well. For this reason alone, as Donald Griffin writes, 'Cognitive ethology presents us with one of the supreme scientific challenges of our times, and it calls for our best efforts of critical and imaginative investigation.'

Cautionary Note

While working on this book my colleagues and I were forced to think about the relationship between ontology (*what* we know about the world) and epistemology (*how* we know what we know). It is this 'how' that creates a lot of problems in the cognitive sciences, for two main reasons. We are trying to understand our own thinking and consciousness by means of our own thinking and consciousness – a circularity guaranteed to confuse. We are also trying to understand animal minds by means of our own minds – an inherently anthropocentric enterprise, as we saw with the MSR test.

Moreover, our wonderful language is a very real problem, as is made clear by Daniel Dennett, one of the most creative and provocative of all the thinkers in the cognitive sciences. Dennett's book, *Consciousness Explained*, is powerful, if perhaps rather extravagantly titled. But his shorter volume, *Kinds of Minds*, is a more focused consideration of animal consciousness. Dennett argues that language may be the 'royal road' to our knowledge of other human minds – with language we defeat philosophical solipsism; with language we know other men and women have minds more or less like our own — but it is a flawed tool for the job of understanding animal minds. We ask questions of the natural world in the only way we can — with our human language — and, in effect, we expect animals to answer with this same language!

In a truly brilliant metaphor, Dennett writes that our language has too much 'resolving power' for this job. He says it is like 'studying poetry through a microscope'. When we think and talk and write about animals and their minds, it *is* like studying poetry through a microscope. The answers will never be totally fulfilling, they may well be misleading and, inevitably, something vital will be lost.

ACKNOWLEDGMENTS

This book would not exist in its present form without the extraordinary editorial contribution of Mike Bryan. Mike is the author or co-author of a dozen books on such widely disparate subjects as golf, baseball, Wall Street, entrepreneurial enterprise and religious faith, who was infinitely patient with this neophyte as I struggled with my first book. He is also a terrific teacher. I might have finished this book eventually without Mike's help but I certainly could not have met some of the publishers' deadlines without Mike. Thank you, Mike, for everything.

My friend and colleague Bill Moyers must also take some of the 'blame'. Bill, who has many books to his credit, strongly encouraged me to write a book and he kindly introduced me to my new colleagues at Doubleday in New York, including Marcy Russoff, assistant publisher, and my editor, senior editor Roger Sholl. Bill was a great inspiration, as were my other public television colleagues, friends and prolific authors, Robert Kotlowitz, Alex Kotlowitz, Robert MacNeil and Jim Lehrer.

Doug Young, my editor at Headline Books in London, and Heather Holden Brown, head of non-fiction, were especially helpful, as were Kelly Davis and Juliana Lessa.

I must thank my employer, 13/WNET, New York's public television station, for management's essential support. So many were helpful but I must single out Bill Baker, our president; Michael

Bessie, board member; Tammy Robinson, head of programming; Carmen Di Rienzo, Esq; Bill Grant, who succeeded me as head of science, natural history and features programmes; Fred Kaufman, executive producer of *Nature*; Janet Hess, science editor; Susane Lee, producer, whose editorial assistance was invaluable; and the entire *Nature* staff, especially Eileen Fraher, the brilliant manager of us all on the *Nature* series.

The generous cooperation of the production team of our television series on animal behaviour at Green Umbrella, Ltd, in Bristol, England, was also essential. To Paul Reddish, executive producer: producers Sanjida O'Connell; Lizzie Greene; Andrew Murray; Rachel Kelly, production manager; and most especially producer Zoe Heron who spent three years as the principal researcher on this project before a frame of film was shot or a word written, I am deeply grateful. The scope and accuracy of Zoe's research was nothing short of prodigious. And, of course, my warm thanks to Peter Jones, the founder and CEO of Green Umbrella and one of the world's most talented producers of natural history films.

My attorney, Robert N. Gold, and my agent, Joe Spieler, deftly handled the business side of this book, for which I am most thankful.

For their cooperation and support I thank, Donald Griffin; Cynthia Moss; Jane Goodall; Professor Nicholas Dodman, DVM, Tufts University; Professors Peter Platt, Barnard, and Joan Ferrante, Columbia University; John Walsh, the World Society for the Prevention of Cruelty to Animals; Elena Araya, the American Society for the Prevention of Cruelty to Animals; Dean Smith, American Antivivisection Society; Patty Bryan; Dr Nick Bryan; and many others who are quoted in this book and for whom its subject is their life's work.

Despite all the fine help I had, I have most probably made some errors anyway. The mistakes in the book are mine alone.

And, of course, I am deeply grateful to my friend, Dr Dennis De Stefano, to whom this book is dedicated, for his patience, encouragement, keen editorial eye and for his ability to put my plight into humorous perspective when I went around asking with great *angst*, 'How did I get myself into this book mess?' Thank you, Dennis.

BIBLIOGRAPHY

Allen, Colin and Bekoff, Marc, *Species of Mind: The Philosophy and Biology of Cognitive ethology*, MIT Press, Cambridge, 1997

Bleibtreu, John, *The Parable of the Beast*, Collier Books, New York, 1968

Budiansky, Stephen, *The Nature of Horses*, New York: The Free Press, 1997

Budiansky, Stephen, *If Lions Could Talk: Animal Intelligence and the Evolution of Consciousness*, New York: The Free Press, 1998

Camhi, Jeffrey, *Neuroethology*, Sinauer Associates, [city??], [date??]

Caras, Roger, *A Perfect Harmony: The Intertwining Lives of Animals and Humans Throughout History*, Simon & Schuster, New York, 1996

Cheney, Dorothy L. and Seyfarth, Robert, *How Monkeys See the World: Inside the Mind of Another Species*, University of Chicago Press, Chicago, 1990

Conniff, Richard, *Spineless Wonders*, Henry Holt & Company, New York, 1996

Coren, Stanley, *The Intelligence of Dogs: Canine Consciousness and Capabilities*, The Free Press, New York, 1994

Darwin, Charles, *Descent of Man* (Great Mind Series), Prometheus Books, 1997

Dawkins, Marian Stamp, *Through Our Eyes Only? The Search for Animal Consciousness*, Oxford University Press, Oxford, 1998

Dennett, Daniel, *Kinds of Minds: Toward an Understanding of Consciousness*, Basic Books, New York, 1996

Dennett, Daniel, *Consciousness Explained*, Little, Brown, Boston, 1991

de Waal, F.M., *Bonobo: The Forgotten Ape*, University of California Press, Berkeley, 1997

Dodman, Nicholas, *The Dog Who Loved Too Much: Tales, Treatments, and the Psychology of Dogs*, Bantam Books, New York, 1996

Dodman, Nicholas, *The Cat Who Cried for Help: Attitudes, Emotions, and the Psychology of Cats*, Bantam Books, New York, 1997

Dukas, Reuven, editor, *Cognitive Ecology: The Evolutionary Ecology of Information Processing and Decision Making*, University of Chicago Press, Chicago, [date??]

Fortey, Richard, *Life: A Natural History of the First Four Billion Years of Life on Earth*, Alfred A. Knopf, New York, 1998

Fouts, Roger with Mills, Stephen Tukel, *Next of Kin: What Chimpanzees Have Taught Me About Who We Are*, William Morrow & Company, New York, 1997

Gazzaniga, Michael, *Mind Matters: How Mind and Brain Interact to Create Our Conscious Lives*, Houghton Mifflin Company, Boston, 1988

Gould, J.L., and Gould, C.G., *The Animal Mind*, Scientific American Library, New York, 1994

Griffin, Donald, R., *Animal Minds*, University of Chicago Press, Chicago, 1992

Griffin, Donald, R., *Animal Thinking*, Harvard University Press, Cambridge, 1984

Hogan, Linda, Metzger, Deena, and Peterson, Brenda, *Intimate Nature, The Bond Between Women and Animals*, Fawcett Columbine, New York, 1998

Knapp, Caroline, *Pack of Two: The Intricate Bond Between People and Dogs*, The Dial Press, New York, 1998

Masson, Jeffrey, *Dogs Never Lie About Love: Reflections on the Emotional World of Dogs*, Crown Publishers, New York, 1997

Masson, Jeffrey and McCarthy, Susan, *When Elephants Weep: The Emotional Lives of Animals*, Delacorte Press, New York, 1995

McFarland, David, editor, *The Oxford Companion to Animal Behaviour*, Oxford University Press, Oxford, 1981

Mitchell, Robert, Thompson, Nicholas and Miles, H. Lyn, editors, *Anthropomorphism, Anecdotes, and Animals*, SUNY Press, Albany, NY, 1997

Moss, Cynthia, *Elephant Memories: Thirteen Years in the Life of an Elephant Family*, William Morrow & Company, New York, 1988

Pinker, Steven, *How the Mind Works*, W. W. Norton, New York, 1997

Premack, David and Premack, Ann James, *The Mind of an Ape*, W.W. Norton, New York, 1983

Restak, Richard, *The Mind*, Bantam Books, New York, 1988

Ristau, Carolyn, editor, *Cognitive Ethology, The Minds of Other Animals*, Erlbaum Associates, Hillsdale, N.J., 1991

Ryden, Hope, *Lilly Pond, Four Years with a Family of Beavers*, Lyons Press, 1997

Savage-Rumbaugh, Sue and Lewin, Roger, *Kanzi, The Ape at the Brink of the Human Mind*, John Wiley & Sons, New York, 1994

Savage-Rumbaugh, Sue, Shanker, Stuart and Taylor, Talbot, *Apes, Language, and the Human Mind*, Oxford University Press, Oxford, 1998

Serpell, James, *In the Company of Animals*, Basil Blackwell, Oxford, 1986

Sheldrake, Rupert, *The Rebirth of Nature: The Greening of Science and God*, Part Street Press, Rochester, Vermont, 1994

Shepard, Paul, *The Others: How Animals Made Us Human*, Island Press/ Shearwater Books, Washington D.C., 1996

Singer, Peter, *Animal Liberation*, Avon Books, New York, 1990

Skinner, B.F., *About Behaviorism*, Vintage Books, New York, 1974

Slater, P.J.B., *An Introduction to Ethology*, Cambridge University Press, Cambridge, 1986

Sorabji, Richard, *Animal Minds and Human Morals: The Origins of the Western Debate*, Cornell University Press, Ithaca, 1993

Sparks, J., *The Discovery of Animal Behavior*, Little, Brown, Boston, 1982

Stevens, Christine, *Animals and Their Legal Rights*, Animal Welfare Institute, Washington, D.C., [date??]

Thomas, Elizabeth Marshall, *The Hidden Life of Dogs*, Pocket Books, New York, 1993

Thorpe, W.H., *The Origins and Rise of Ethology*, Heinemann Educational Books, London 1979

Vauclair, Jacques, *Animal Cognition: An Introduction to Modern Comparative Psychology*, Harvard University Press, Cambridge, Mass., 1996

Wilson, E.O., *Consilience: The Unity of Knowledge*, Knopf, New York, 1998

Wilson, E.O., *Sociobiology*, The Abridged Edition, Harvard University Press, Cambridge, 1980

INDEX